Answers For Everything

Answers For Everything

Joe Paladin

iUniverse, Inc.
New York Lincoln Shanghai

Answers For Everything

iUniverse, Inc.

For information address:
iUniverse, Inc.
2021 Pine Lake Road, Suite 100
Lincoln, NE 68512
www.iuniverse.com

ISBN: 0-595-28071-4 (pbk)
ISBN: 0-595-65731-1 (cloth)

Printed in the United States of America

This book is dedicated to everyone who has searched for truth in the past contributing to what is known today, and to everyone who searches for truth today. I hope that one day each individual becomes their own philosopher, only then can mankind be free of the ignorance that has plagued society throughout history.

Contents

1

Introduction

Hi. My name is Joe Paladin. I would like to present answers for the bigger questions about life and creation. Sometimes the world around us, even one's personal role in things, can appear complex. Things appear complex when the underlying truth of things is not completely understood. Truth is often the simplest solution, not the most complex.

Not just any simple solution will do. Solutions are true because they represent realities of life. Truth is truth because it drives events all around us, not because a particular individual claims something to be truth. Truth is verifiable. Even if not a single person understands the truth of this universe, laws of creation and life still exist. Truth can be found, and some truths are more valuable than others. For example, if someone throws coins into a fountain a truth exists that names the number of coins in the fountain. That truth is less valuable than the truth Einstein discovered demonstrating the equivalence of mass and energy. That truth Einstein found is more important because it put the power to either annihilate the planet, or build ships that run for decades without refueling, into the hands of mankind. Whether that is good or bad is beside the point, clearly some truths are more important than others. As powerful as the truth Einstein found is, an even greater truth exists. That truth is what this book is about.

An individual takes the first step on the path to knowledge by acknowledging personal ignorance. For the most part, mankind has done well in that respect. All around the world scientists and researchers freely admit that there is much more to the universe than people currently understand. Acknowledging personal ignorance does not mean that personal experience in meaningless, on the contrary, personal experience is valuable. Personal experience allows an individual to evaluate information and products. An open mind, combined with experience and a willingness to test new ideas or products, is the path to truth.

Say you had a lot of friends who believed that the world around them was destined to face catastrophic changes some time in the future. What if these friends where taking no action, not knowing exactly when or what the exact nature of those changes would be?

Say these friends based their belief on faith and various interpretations of the Bible. Let us consider two types of faith. Faith can be trust without detailed knowledge. Faith can also be trust that includes detailed knowledge of events that will come to pass, and why these events will happen. Here is an example.

Today's soldier has faith in his weapon. The soldier knows that gunpowder explodes and propels a bullet. This is faith combined with knowledge. It would take a different kind of faith for a Greek Mercenary who never saw a gun to take that weapon into battle against overwhelming numbers of swordsmen and archers.

The faith of the mercenary would have to be trust without knowledge if he were simply told that he must look down the sight and pull the trigger. In the face of overwhelming odds, without knowing about guns even via second hand information, that would be simply faith.

Christ told people 2000 years ago that mankind would face certain global challenges, and that people must prepare to survive these challenges. Christ painted the best picture possible at the time, covering a few natural aspects of the challenges that would be faced. Revelation provided additional information. The language during the days of Christ influenced how much knowledge could be passed down in the Bible.

People were asked to prepare for these events based on faith that could not include detailed knowledge at that time in history because of the low level understanding regarding technology and nature during those times. Christ mentioned economic expansion and return on investment in a parable. During the past 2000 years technology has progressed rapidly, and economic expansion has occurred. The Bible is a book of higher truth. That truth is directly related to physical technology far beyond anything mankind achieved in the 20th century. That truth is also related to the spiritual reality of mankind, a reality that has been poorly understood even by the best Biblical scholars. The details of this truth will be presented.

Today it requires two physics breakthroughs to convey the knowledge behind the physical events predicted in Revelation. Fortunately, those breakthroughs are easy to visualize. Most people will understand these things very well. Technology has advanced such that all the building blocks of understanding are in place.

The first breakthrough is a grand unification of natural forces. The second is a completion and simplification of Einstein's work.

Pretend that you yourself made a very simple breakthrough in physics that explained the exact nature of the events described in Scripture. Now, pretend that only you and one other inspired physics researcher knew about this breakthrough.

Knowing the details about a catastrophe threatening all of mankind would be a great burden to bear. The burden would be even greater knowing that mankind is technologically capable of preparing for the future, but is not doing so.

Pretending that you have made such a breakthrough, you would probably want to be 100 percent sure of what you had discovered before you told anyone. You would need to be able to demonstrate this truth to anyone on his or her own terms if you really expected an entire nation to take action.

If the threat were an asteroid hurtling through space certainly people would do something. People understand being hit by a rock. Unfortunately, the events described in Revelation are not that simple. For example, if an advanced society called Atlantis did actually exist, and if it was destroyed in days by natural events as stories describe, and if even the people of Atlantis with space travel technology were caught of guard, could it be that mankind today is missing something as well?

Now pretend you had time on your hands and decided to spend as much time as it might take to disprove your new discovery. In particular, you try to disprove the part about catastrophic change occurring within your lifetime. Say that after months of research gathering all the empirical data you could find everything you found further confirmed the new model. Tears might roll down your face for a number of days, realizing that people knew not what they faced, and also knowing that people tended to wait until after the catastrophe had hit to do anything about it. In this case, that waiting without preparing for physical survival would result in the physical destruction of mankind.

So, you decide to present this new physics model that defines catastrophic planetary change to a variety of people to see what they think about it. The idea being that if most of them didn't agree with it, or couldn't understand it, then you could say that there really wasn't much you could do about it anyway. Trying to help people that are convinced they don't need any help is not possible. A survey that indicated people just could not understand or agree with such a thing would take some of the weight of the world off your shoulders.

Say everyone that you talked to understood the discovery, and they generally agreed with your conclusions regarding the future of this planet. If you didn't have millions of dollars in the bank to launch some kind of campaign, and you didn't have fame to reach out to the millions of people who would have to be reached, you might face a moment of frustration.

That is the position I found myself in a few years ago. I achieved the highest plateau of elation recognizing and understanding all of creation. From there I descended into the deepest catharsis I have ever experienced looking at the people around me. When I got over that I decided to make some plans.

That time in my life is close to when the story of discovery ends, and the writing of this book begins. You can imagine that it would be challenging to write a book that is intended to teach most people about everything. Most people do not have an engineering or scientific background. The reason why physical creation has not been understood in the past is that life is the key to creation. There is more to it than writing some new math formulas. Fortunately, a degree in physics is not necessary to understand these things. A technical background does help; it just isn't absolutely necessary unless an individual wants to design some new products.

Each chapter of this book builds understanding for future chapters. This first chapter is an introduction that walks through key life events that lead to the conclusions presented latter. After building up to an understanding of creation and life I will present solutions and predictions for the future.

Growing up

I grew up in a Christian family with 4 younger brothers. My family lived in a house in Southbridge, Massachusetts. My Dad was an accountant. My mom had a big garden in the back yard, so we had fresh fruits and vegetables. Between playing chase and dodge ball in the summer, sometimes my brothers and I would go out and pick blackberries. My mom made the best blackberry pie around. Life was simple and good.

I have been interested in technology, philosophy, and religion as early as I can remember. My first energy experiment involved seeing whether or not the family sedan could run on sand. This experiment was unsuccessful. The result was a large mechanic bill to clean out the fuel system and replace the carburetor. My father was not happy. I decided it was best not to mess with my parent's stuff.

When I was about seven years old I decided to read some of the Bible. I read most of the New Testament and all of Revelation. After reading Revelation I recall gazing out past my front yard envisioning a global catastrophe that this world would face in the course of my lifetime. I had a strong conviction that I would lead people through these times. Ten seconds later I said to myself: "Wait a minute, there's about 4 billion people on this planet! What are the chances of me being responsible for leading people?"

What struck me was a sense of personal responsibility. Being seven years old, there wasn't much I could do. I decided to just live my life. If I were meant to play a particular role in the future that role would have to find me.

I attended Southbridge Elementary School. I did quite well in math after getting over a dyslexia problem. Dyslexia involves reversing numbers or letters. I would correctly solve a math problem in my head, getting an answer of 51 for example, and then write down 15 on the test. I spent some months with a special education teacher who had me do some exercises. I would look at a long sequence of numbers, and then write them down from memory. That fixed the dyslexia, since I realized what I was doing wrong, and I could correct it. Some languages of the past read from right to left, backwards essentially. Could that have something to do with dyslexia? In any case, dyslexia is easy to correct with a little attention.

College

For my first few years in high school I was planning on applying to MIT. I had a high enough GPA and SAT scores to qualify for MIT. During my junior year I decided to shop around for colleges. I liked Worcester Polytechnic Institute (WPI) and decided to attend a summer camp at WPI. I was impressed with the practical nature of WPI, and the focus on teaching the social impacts of technology in addition to technology. WPI was rated one of the top schools in the Northeast. I decided that was where I wanted to go. I applied for early admission and I was accepted.

I enrolled as an Electrical Engineering major. During my freshman year I took my first college physics class. During a lecture on the behavior of charged particles I asked the professor what was between the two charged particles that caused the particles to experience a mutual force. If there was a complete void between the particles, how could one affect the other? My physics professor was caught a little off guard. He quickly recovered and told the class that was the way God made it.

At the time, I wasn't very satisfied with the "God made it that way" answer. I had learned that while attending Christian Doctrine classes, which were free. Here I was paying $20,000 per year for a technical education.

Today, I commend the honesty of that physics professor's answer. It is true that God made it that way. More importantly, my professor defined the limits of mankind's understanding of physics. There are only a handful of very good physics department heads in the world. At the very least, these people can say what is known and what is not known in the field of physics. I am sure that in the past cave men asked why fire burned, and a wise cave man said God made it that way.

There is a precise answer to that question about particle interaction. That answer is key to just about every advanced technology people have dreamed of but have been unable to build. The way to solve a mystery is to gather information and take a very close look at anything that does not make complete sense. Scientists freely admit that gravity is not well understood. If it were, people would be building vehicles with gravity propulsion systems. That is not the only hole in mankind's 20th century understanding of physics.

Einstein said that the magnetosphere was the greatest mystery in physics. Einstein was brilliant. He made original advances. Unfortunately, most people who came after Einstein studied Einstein rather than doing their own research.

People overlooked the weakness that Einstein himself saw in his own work. A few less than honorable researchers began to invent things to sweep things under the table, like that observation regarding the magnetosphere. A rotating iron core model may satisfy the uneducated public, but it does not satisfy people who really know what they are talking about. If Einstein considered the magnetosphere the greatest mystery in physics, do you think he would have missed a trivial conducting iron core model? I don't think so.

A rotating iron core model cannot explain the actual behavior of Earth's magnetosphere. Scientists know that iron loses its magnetism at core temperatures. A large flowing mass of magma would have inertia, which would preclude very rapid accelerations and decelerations. The entire magnetosphere can change dramatically in minutes during magnetic jerks or magnetic storms. Not only that, the entire magnetosphere completely reverses. Even a grade school student knows that an object in motion tends to stay in motion unless acting on by an external force. So, a model that says that the magnetosphere is caused by a huge rotating core of magma, and that for no particular reason the entire core of magma starts spinning in the opposite direction is ludicrous. Although many other points can be made to demonstrate an iron core model is false, one valid point is sufficient to disprove it. It doesn't take an Einstein to recognize that something is false even if one does not know what the actual truth is.

I decided to switch to Mechanical Engineering. I figured that even if the Electrical Engineering guys could build electrical devices they lacked a basic understanding of what they were dealing with when they talked about electrons.

College was a blast. I made a lot of new friends and had a great time going to fraternity parties at Theta Chi. My friend Mark Salsman was the leader of the

Theta Chi stud patrol. Since the ratio of guys to girls at WPI was about 7 to 1, it was necessary to go out and visit other colleges in the area, and invite girls to the fraternity parties. Mark would go visit Holy Cross, Assumption College, Worcester State, and other schools in the area.

Towards the end of my freshman year at WPI I headed out to Daytona Beach on a Spring Break road trip. That was a wild trip. I traveled with a crew consisting of two Theta Chi brothers and two other friends. Mark Salsman and John Ashley were the two Theta Chi brothers. Paul Zaloom, David Creed, and myself made up the rest of the crew. We all worked out together at a gym in town.

We drove to Daytona Beach in two cars. Paul had an older Impala and Mark brought his Toyota. It was amazing that Toyota reached 100 miles per hour with three people in it plus baggage. We had CB's and call signs, switching drivers now and then to make the best time. My call sign was "Driver 8". R.E.M. was playing on the radio when I was driving. We managed to arrive in Daytona with only 1 speeding ticket.

After spending a few days at Daytona Beach we planned to travel down the coast and hit Fort Lauderdale. Before leaving Daytona I found out my parents had called. The crew and I were visiting a cousin of mine living in the Daytona area before leaving for Fort Lauderdale. My cousin told me to call home as soon as I could.

I called to see what was up. I was finishing my freshman year and as it turned out the scholarships I received were all 1-year awards. My parents wanted to talk about money for my next three years at WPI. I was doing well on my courses and making good grades.

I had been considering Air Force ROTC. My Mom and Dad told me they had visited WPI's Army ROTC office and talked with the Colonel in charge. Based on my admission information, the Colonel thought I could qualify for a full scholarship covering my next three years of tuition, plus a monthly stipend for general living expenses. There was only one of these command-sponsored scholarships available, and it was awarded by the Colonel in charge of the regional ROTC office.

Being on Spring Break I was more interested in going to clubs and increasing my holdings of spring break T-shirts. I said I would think about it.

When I got back I did the application for the Army scholarship. I scored the maximum possible points on the Army physical fitness test. I made a good impression and won the scholarship.

ROTC is flexible. The program accommodates existing class schedules and exam preparations. The scholarship required maintaining a GPA better than 3.0, which was no problem. I graduated from WPI three years later with honors.

My original intent was to serve in the Reserves after school. Some people enrolled in the ROTC program at WPI were requesting active duty service and being turned down in favor of West Point candidates. The ROTC program at WPI served about 10 different colleges in the Worcester area. Active duty assignments in the officer corps are competitive. I requested Reserve Forces duty, expecting that was a sure thing with all those people requesting active duty.

Apparently Uncle Sam wanted me in particular on active duty. I received orders to report to Engineer Officer Basic Course (OBC) the following October. After that, I would move on to my unit in Germany after completing training. That was fine with me. I decided to make the best of my active duty tour and gain as much experience as I could.

It was April of 1990. My courses were complete, I had a couple weeks before graduation, and I had six months before my October reporting date.

All WPI grads do an interactive qualifying project (IQP). The purpose of the IQP is to do a real project applying the engineering skills that have been learned to solving a problem in society. My IQP was alternate energy, photovoltaics and local energy storage systems in particular.

My friend Avi was in charge of managing the construction of WPI's solar car entry for the GM Sunrayce. The race was being held during the summer after graduation. Avi was having trouble finding enough people who would be available to finish building the car and be part of the race team. Most graduating seniors were starting new jobs during the summer.

Thirty engineering schools from around the country entered the GM Sunrayce. The top three cars would receive a grant and go to a worldwide race sponsored by GM in Australia. MIT was one of the schools that entered the race. Avi was determined to at least beat the MIT car if not win the entire race. The course would begin in the Disney World theme park in Florida then run cross-country to Detroit, Michigan. The course ran on secondary roads since the cars were not expected to run at highway speeds. Our team's car could reach 60 mph running with batteries, but it was more efficient to run slower.

I decided to join the race team and help finish building the car. Our team bought a low cost trailer and towed our solar car to Florida. The speed qualification for poll position was held at the Daytona 500 racetrack. The car also had to meet certain entry requirements for the race. Requirements included size limitations for the solar array, functional brakes, etc.

The batteries we used were the best lithium ion batteries available. The batteries alone cost nearly $100,000. Avi managed to get the batteries donated to WPI's solar car project. We also had one of the most efficient motors available.

We placed second in the speed test out of 30 cars in the race. All the cars lined up in the Disney World theme park. We rolled out of Disney at a good pace on a sunny day using both power from the sun and battery power. We took the lead on day 1. Unfortunately, after the first day's run, we noticed a larger battery drain rate than we had originally calculated. It took us three hours of racing the next day to realize that the brake calipers were rubbing against the rotors. We finished the car only a day before leaving Worcester so we had only been able to test it in the parking lot.

The challenge of building a solar car includes reducing the weight and minimizing the frontal cross section to reduce air drag. We had to really cut things close with materials. The body was completely composites, and the mounting points were not very stiff at higher speeds or over bumps on rough surfaces. This body flex caused problems with the active braking system. Otherwise we would have had enough clearance. The car's regenerative braking system stopped the car almost as fast as calipers could. Regenerative braking involves engaging the electric motor as a generator during braking so that power flows back into the battery system. We backed off the calipers and continued the race.

On the third day of the race we encountered a bureaucratic racing nightmare.

Rather than allowing us to recharge the car batteries throughout the day using the photovoltaic array, race officials impounded the cars at the end of the race day. We were only allowed an hour of charging time early in the morning before the start of the race. There was a race window of about 6 hours during which a car could log race miles. Each day was a race stage. The car with the lowest cumulative time would win.

If we ran faster using batteries we would finish the stage earlier but would not be able to get much charge back into the batteries. The impound area was under a tent in the shade. Race officials were concerned that people would cheat and run an extension cord out to a racecar. This took away our team's main advantage of running at higher speeds with battery power plus solar power.

Some cars in the race ran slower but more efficiently using little more than onboard solar array power. These cars relied much less on batteries, conserving weight. The University of Michigan was racing such a vehicle.

That wasn't the only thing the University of Michigan had going for it. The University of Michigan's race team leader was from their school of business. She was a very attractive girl. She was also an effective fundraiser. She contacted all the major automakers and many related companies in Michigan.

Michigan arrived at the race with two semi-trailers containing complete machine shops. Mounted on top of the trailer was a real time satellite tracking system capable of locating individual clouds, and then downloading accelerations and decelerations to the Michigan solar car. They could keep the car in the sunshine racing right behind a cloud if necessary. The Michigan car had nearly $1 million dollars of the best solar cells available, and each one was applied individually to maximize area within the allowed envelope. It was an unfair advantage, we wondered if they new the judges and had anything to do with the last minute impounding rule.

Racing down the road we saw a solar car in a ditch on the side of the road ahead of us. Approaching closer, we noticed it was the MIT car. The nose of the vehicle was on the ground. We learned latter that MIT had lost an axle after hitting a pothole. The MIT crew had under designed that component and they didn't have a spare to replace it. That was sad. We waved encouraging signals out the back window of the chase vehicle and raced on.

We managed to win a few race days. We scored first place in the hill climbing competition. They day before the hill climb we stopped on the side of the road to charge the batteries, and took the hit on that days stage. We figured there was no way to catch the Michigan car so we decided to win as many prizes as we could. We already had T-Shirts, and we got special GM Sunrayce Swiss Army knives for the hill-climbing day. About 12 cars finished the race. We finished middle of the remaining pack so we didn't get to race in Australia.

It was a fun race. Every town we raced through had crowds of people lining the streets to watching the solar cars race by. The average speed was about 20 miles per hour so it wasn't like watching the Indy 500. The GM Sunrayce vehicles were unique though, and certainly energy efficient.

Computer Aided Design

After the race I returned to Worcester, Massachusetts. It was the end of June so I still had almost four months before going out to OBC in Missouri. During my Junior and Senior years I had been doing some part time work for Cadkey Incorporated.

Cadkey was the first three dimensional Computer Aided Design (CAD) platform. Cadkey was three dimensional long before AutoCAD. Cadkey has a built-in Computer Aided Design Language called CADL. This language allowed programs to be written to generated 3D geometry and precise 3D animations.

I loved doing 3D designs and CADL programming. During my junior year my CAD professor, Bob Bean, decided to refer me to the people at Cadkey. I was introduced to Cadkey's president, Peter Smith. Peter asked me if I wanted to work at Cadkey during the summer doing software testing. I said yes and spent some time commuting to Connecticut that summer.

After doing Cadkey beta testing for a few weeks Peter asked me if I wanted to help with a special project writing software to generate three dimensional surfaces and solids. Solid modeling was a new concept back then. I wrote some code that generated variable cross sectional sweeps in 3D space using CADL.

A Cadkey customer named Dr. Jarvik lived in New York. Dr. Jarvik designed the first artificial heart. At that time, Dr. Jarvik was using Cadkey to design his new Jarvik 7 artificial heart. Peter wanted to support Dr. Jarvik's effort, as it was good publicity for the company.

I ended up taking a couple trips to New York to visit Dr. Jarvik and help out as best I could. It turned out that Dr. Jarvik was doing quite well and didn't need much help. I enjoyed the visits. This experience led to a job offer that I accepted prior to going out to OBC.

Bob Bean was starting his own company developing a Cadkey add on package call Draftpak. I had been doing some part time work for Bob during school helping to create the first release of Draftpak. Bob incorporated a company called Bay State Technologies to sell Draftpak. Bob asked me if I wanted to work for him full time until I started OBC, and that kept me busy until October. Bob also wanted me to continue part time will I was on active duty. Software programming can be done at home as easily as in the office. I didn't know what my schedule would be on active duty so I could not commit to that. Also, I didn't really need the money, and I did want to have some free time.

The Army

When October rolled around I was in peak physical condition expecting some difficult training at OBC located in Fort Leonard Wood Missouri. We did actually do some tough training. We were up at 6 AM every duty day for physical training (PT). It wasn't as tough as I expected. When I arrived for in processing I found out that I was going to receive a $5,000 temporary duty assignment (TDY) advance. This was in addition to my regular pay as a second lieutenant. The TDY advance served to cover food and housing expenses away from a permanent duty station. Rather than staying in barracks most new officers including myself paid for officer's quarters on base. That was a big change from ROTC. Now that I was a commissioned officer things were better than being a cadet.

Engineer OBC was a 5-month course covering things like demolitions, bridge building, earth works, road building, and cement structures. The training included field exercises. We exchanged our comfortable apartments for tents in the field and practiced using live demolitions, building bridges, throwing live hand grenades, and qualifying on machine guns and M16's. It was a fun course.

Compared to engineering courses at WPI, OBC tests were not very difficult. I really didn't need to study in the evenings when we were not in the field. I spent a lot of time playing pool in the officers club and going on an occasional weekend road trip.

Three months after I arrived at Fort Leonard Wood the U.S. launched Desert Storm in response to Iraq's invasion of Kuwait. We were at war. Considering America had not been involved in a major military action for many decades, I wondered about the chances of this occurring not long after my arrival on active duty.

Now that the country was at war on the other side of the world security tightened at Fort Leonard Wood. The unit that I was assigned to after OBC was based in Germany and preparing to deploy to the Gulf. There was a real concern over the use of chemical or biological weapons against U.S. troops. The army was inoculated.

The array of preventative shots we all received at OBC read like something out of a poem by Edgar Allen Poe. Anthrax, Plague, Bubonic Plague, Black Death, Small Pox—a list that might be carried by the four horseman. Building up antigens to biological agents can't hurt though, so I didn't mind the shots.

During OBC new officers were allowed to request additional training prior to deployment. The Army has a variety of special schools like Ranger School, Airborne School, and Scout Platoon Leader Course. Not everyone gets to attend these schools. Towards the end of OBC I found out that I had qualified for two open slots to attend special teams training. I was given a choice of Airborne School or Scout Platoon Leader Course (SPLC).

I did sport parachuting at WPI, so I do like jumping out of airplanes. Airborne insertion is not a primary mission of the Corps of Engineers. It is possible that a team of engineers could be attached to an airborne unit.

I decided that I really didn't need more training to jump out of a plane and pull a ripcord. On the other hand, SPLC would help me with tactics and terrain navigation. Scouting out land and enemy positions would be valuable training.

Terrain knowledge is key for any army officer. I selected SPLC and received a set of orders bound for Fort Knox, Kentucky.

SPLC was a challenging school. Out in the field we averaged less than a couple hours sleep per day. We spent the nights on missions navigating in the dark with vehicles and on foot. Desert Storm had been in progress for a few months at that point. Commanders in the field decided to rotate some experienced sergeants back to the States to help train people preparing for deployment to the Gulf. Some of these Sergeants were tasked to help train my SPLC class at Fort Knox.

We learned many different things from those sergeants about what combat in a desert environment really involved. For example, the sand in the desert is very hard on vehicles. The air intakes on the Bradley Fighting Vehicles and M1 tanks needed secondary air filters to prevent the main filters from becoming blocked too quickly. People were writing home asking for panty hose to be sent over to stretch over intakes as a filter. Desert Storm was the first military action for many new weapon systems fielded by the Armed Forces.

The M1A1 Abrams Main Battle Tank proved to be a dominant force on the battlefield. Sergeant Smith was a tank driver who described his experience as part of the main force moving across the desert driving back Iraqi forces. Some tanks are equipped with mine clearing equipment, which consists of huge cement rollers pushed along in front of the tank. The rollers are combined with an electromagnetic field generator designed to set off electromagnetic mines out in front of the tank. The rollers simply depress and explode regular mechanical mines.

The army has limited numbers of mine clearing equipment sets assigned to each unit. Not every tank could be equipped. These large rollers slow down the tank, and make the tank an easier target for weapon systems other than mines. One tactic used involves a lead vehicle equipped with rollers while other vehicles follow the path that has been cleared. The problem with that tactic is that it concentrates forces and makes it difficult to form a wide front to cover open terrain. In Iraq a wide front was needed, and it was not always possible to locate Iraq minefields in advance. Clear an entire dessert that might contain a mine at any location is not feasible. Commanders had to do their best and direct routes of advance avoiding known mine fields. Often an armor unit did not know they had entered a minefield until a mine was detonated.

Clearing minefields is a primary mission of the Corps of Engineers. Naturally the Corps maintains their own mine clearing assets. For example, the Corps of Engineers has a Mine Clearing Line Charge (MICLIC) that launches a small rocket pulling a long string of C4 charges through the air behind it. The rocket lands laying out the charges behind it to detonate in a line clearing a path for vehicles. Unfortunately, a MICLIC is not very helpful after a vehicle has already driven into the middle of a minefield.

Sergeant Smith's tank hit a mine that destroyed the right tread of the M1 leaving his tank immobile. The battle does not stop because one vehicle is hit. Units to the left and right have to cover a larger field of fire and the front drives on. Sergeant Smith's tank commander called in a situational report (Sit Rep). All the crew could do was wait until help could arrive from the rear. Air and land rescue units are held in reserve prepared to deploy where they are needed, but that takes time especially if a tank recovery vehicle is needed. Before help could arrive three Iraqi tanks approached the M1. It is much more difficult to hit an M1 tank moving 50 miles an hour across the desert than it is to hit a stationary M1 in the open. Iraqi forces were being destroyed on sight as the coalition front advanced. Apparently these Iraqi units had managed to stay out of sight.

The M1 was not about to drive anywhere soon. The crew did not want to walk away from the tank in the middle of a minefield and expose themselves to small arms fire. The M1 turret still worked, along with the targeting system. Sergeant Smith's gunner destroyed the first enemy tank immediately after it crested a berm to their front. As he set his sights on the next target the whole tank was rocked and their ears were ringing. They had been hit. Fortunately, the M1 is equipped with depleted uranium armor, which is the toughest material on earth. The M1 armor plates are set at angle to deflect some of the energy of incoming rounds. The M1 is also designed low to the ground to make it an even more difficult target. The round bounced off the M1 ringing the ears of everyone in the tank.

The M1 took two more hits before destroying the remaining two Iraqi tanks. An M88 recovery vehicle arrived from the rear several hours latter. An M88 is a tank almost twice the height of the M1 with a large boom pivoting at the back of the tank and supported at the center. The M88 has a large winch and cable system capable of pulling another tank completely out of a ditch. The M88 recovered the M1 and brought the tank and entire crew safely back to friendly lines.

Half way through my training at Fort Knox the unit that I had been assigned to had already deployed to the Gulf. They were at 120 percent strength, which meant that all positions were full and they had extra personnel. Usually extra personnel, especially officers, end up getting assigned to the battalion Command Post and sent down to companies as necessary.

I wanted my own platoon for my first assignment. I did not want to sit around a Command Post (CP) being the lowest ranking officer around with very little responsibility.

Before I completed SPLC my commander asked for some volunteers. He had orders to provide additional personnel to support a buildup of forces in South Korea. At the time there was some concern that the conflict in Iraq could cause North Korea to take advantage of the U.S. concentration of forces in a different part of the world. There were also concerns about North Korea building nuclear weapons.

The tactical situation in Korea is much different than the Middle East. Korea is mountainous terrain with cold winters. Terrain can provide an enormous advantage if used correctly. Terrain makes it possible to confront superior forces that have both numerical and weaponry advantages.

Mountain ranges divide North and South Korea. The North Koreans have had decades to fortify that dividing line. There has never been an official truce between North and South Korea, so conflicts still arise on occasion in the Demilitarized Zone (DMZ). The DMZ separates North and South Korea. This buffer zone is filled with mines laid mostly by the U.S. and South Korea.

North Korea holds the high ground that is North of the DMZ. South Korea holds the lower ground, South of the DMZ, which is mostly rice paddies and hilly terrain. Up until the late 80's U.S. forces worked together with South Korean forces patrolling the DMZ. Now that mission is completely in the hands of South Korea.

North Korea maintains artillery emplacements in the mountains north of the DMZ. These guns are dug into the mountains and ride on rails. The gun rolls down with gravity, opening a huge steel door just as the gun fires. The recoil drives the gun safely back into the mountain ready to be reloaded. The North Korean artillery is capable of hurling a 1000-pound shell 20 miles or more into South Korea.

Many of the U.S. forces stationed in Korea, along with a couple major cities in South Korea, are within range of these guns. Ground forces within ten miles of the DMZ have little chance of stopping a first wave assault if one were to occur. Patriot Missiles cannot shoot artillery shells out of the sky. Politically, the U.S. maintains ground forces in Korea to deter North Korean aggression. Presumably, North Korea would not attack a small vulnerable U.S. force knowing that the rest of the U.S. is behind that force. That reasoning may have held true many decades ago. In today's world it is best to consider deployments more carefully realizing that the enemy is not always rational or interested in self-preservation.

My commander at Fort Knox told us that a number of combat engineer platoon leader positions needed to be filled in Korea. I volunteered. At the time I didn't know much about the tactical situation in Korea. I knew that they needed platoon leaders over there and I didn't get the impression I was needed in the Gulf. I packed my bags shortly after Scout Platoon Leader Course and flew out to Korea on a commercial jet.

I arrived in Korea looking around at everything written in a foreign language. I had a Korean—English dictionary with me but Korean is not an easy language to just pick up. Fortunately, my orders preceded me. A driver arrived from Camp Mercer waving a sign at the airport to pick me up and bring me in from Seoul. Camp Mercer would be my home for the next year.

On the drive back I heard some stories about the wilder days of Camp Mercer. Things were less exciting now that the new commander did not allow combat engineer dump trucks to be used to transport groups of local girls to parties on base. This hardship duty station was changing in the wake of heightened security. I decided to make the best of it while I was there.

Arriving at Camp Mercer I walked into a compound not much bigger than two football fields long surrounded by high fences and double rows of barbed wire. Officer quarters consisted of a small room about 12 feet long and 8 feet wide. It was possible to rent an apartment off base, but it was discouraged. It would have been inconvenient traveling in every day for PT.

On my second day I walked down to the motor pool to take charge of my new platoon. My Combat Engineer Platoon consisted of four old dump trucks manufactured in the 1950's. These dump trucks would certainly spark fear in the eyes of advancing North Korean troops! Maybe we could engineer oil slicks to impede the advance of armored vehicles, considering the number of class 1 and class 2 leaks I saw on the maintenance reports. Units in Korea are low priority for spare parts, possible they didn't make parts for those dump trucks anymore.

The dump trucks were tactical vehicles capable of hauling loads of mines up to the DMZ. No new mines had been laid for 20 or 30 years. Those minefields were now just as dangerous to enter for South Korean forces as for North Korean forces. The mines themselves were decaying.

I began to settle into the daily life of Camp Mercer. Not long after I arrived I decided to check out the local nightlife with a couple other officers on base. The first club we walked into I sat down and an attractive Korean girl walked over, sat on my lap, and asked if I would buy her a drink. I was single at the time, though I had been dating 1 girl I met in Missouri. I bought some drinks.

I learned later that it wasn't all about my personal charm and excellent physical condition. Some bars hired girls to get people to buy them top shelf drinks. These drinks are expensive, and they are served watered down to the girls. An unsuspecting GI could find himself being drunk under the table doing shots, never knowing what hit him. It was entertainment.

It wasn't long after I arrived in Korea before my company deployed north towards the DMZ for field exercises. I led my convoy of dump trucks north, looking out the window of my Humvee to make sure no one got lost or broke down on the side of the road. The problem with Korea is that there is no side of the road. They build the front walls of buildings right up to the road. Outside densely populated areas the roads are all narrow dirt roads cut into the hill on one side with a ditch on the other. Traffic jams are almost unavoidable without breakdown lanes.

My first mission was to run a demolitions range near the DMZ. Demo training was my favorite part of OBC.

A primary mission of engineers in Korea is the destruction of bridges to impede a North Korean advance. After training the previous day on cratering charges, I planned to train the men on steel cutting and breaching obstacles. A number of large steel I-beams were available similar to those used to support bridges. Cutting steel I-beams with C4 involves properly placing charges such that they all detonate simultaneously around the I-beam. This prevents the explosion of one charge from blowing the other charges off the I-beam.

In addition to steel cutting, we set up some bangalore torpedoes. Bangalore Torpedoes are thin metal tubes filled with C4. Bangalore sections are connected and pushed under wire obstacles to clear the wire. My squad leaders went to work setting steel cutting charges and running Bangalore Torpedoes.

I took out my laminated demolition book to calculate safe distance on the range. I added up how many pounds of C4 we were using, which was something like 30 pounds or so, I don't recall exactly how much. Dropping down the chart for pounds of C4, then across for minimum safe distance, I calculated 300 feet or about 1 football field.

There was a berm on the range about 300 feet back that was a little over waist high. Three hundred feet was the minimum safe distance, and it looked like other units had used this position previously. Around that time my company commander arrived in his Humvee to inspect the range and watch the result of my platoon's steel cutting charge placements. Although it is recommended that one exceed the safe distance, it is very tempting to get a good view of the explosion.

We used detonating cord (det cord) to link all the charge placements for simultaneous detonation, and timing fuse to initiate the explosion and give us enough time to move back to the berm. Det cord is a thick wire with a core of C4 that comes on 100-foot spools. Det cord can be run in lines and tied to other explosives so that a single blasting cap can be used to detonate a small section of det cord. When that small section detonates everything else attached to the det cord also detonates. Sgt Proteau lit the timing fuse, and his squad along with my platoon sergeant moved back to the berm to join the rest of the platoon. Everyone was behind the berm with time to spare. The explosion produced a large cloud of dust down range. I watched as a section of steel I-beam hurtled through the air towards my platoon. Fortunately, it passed over everyone's head. I started to wonder about the accuracy of those safe distance charts.

My commander, Captain Johnson, walked over to me balanced between anger and a sheer sense of relief that no one had been hit. He wanted to know if I had calculated the safe distance on the range. I showed him the chart that clearly indicated a safe distance of 300 feet. Good thing we were not a little farther away.

Captain Johnson quickly pointed to footnote 5 in very small print at the bottom of the page. Footnote 5 said that the safe distance for steel cutting was a minimum of 1000 feet. I was tempted to ask why that wasn't in large bold letters at the top of the page. That would have been an excuse though, which only makes things worse. I acknowledged how serious the situation could have been and promised not to make that mistake again.

In truth, we were all surprised that a big piece of I-beam like that could have been launched so far from the blast site. I recalled Sgt. Proteau and his squad moving a large rock over towards the I-beam. The rock may have been positioned such that it launched half the I-Beam with the other half being braced against the big rock. A classic example of the Newtonian observation that every action has an equal and opposite reaction. The trajectory of the I-beam wasn't actually right over our heads; it was off to the left some. Squad leaders can be mischievous. I had wondered why Sgt Proteau was so enthusiastic about moving that big rock with his squad. I suppose it did serve to simulate a rock or concrete bridge abutment near the charge placements. We all learned something about steel cutting that day.

The next day my platoon moved off the range back into a staging area. I was expecting an uneventful day when my driver walked over to tell me the platoon sergeant needed to see me. He mentioned something about Sgt. Proteau driving a truck off a cliff. This sounded a little extreme to me. Sgt Proteau was licensed on more equipment than anyone in the company, which is why he was tasked out to help head quarter's platoon.

Sgt. Proteau had been tasked out as a driver to support digging-in some 1st Infantry Division fighting positions. Proteau was driving a flat bed truck with a bucket loader chained to the bed. A bucket loader is used to scoop up dirt and load it into a dump truck. It is the same as the civilian version, but with Olive Drab (OD) green paint.

Sgt. Gales, my Platoon Sergeant, brought me up to speed on the situation. Sgt. Proteau was driving the truck down a steep mountain road. The brakes were not able to handle the weight of the trailer and bucket loader so the truck gained speed going down the mountain. He tried desperately to slow the vehicle down using the emergency brake and shifting into lower gears but he lost control. Proteau and the other soldier riding with him managed to get out of the cab. The flat bed was off the edge of the road and stuck.

Sgt. Gales said they sounded scared. I got the grid location from Sgt. Gales and drove out there in my Humvee. On the way over I wondered if they were exaggerating things. That was not the case. Half of the flat bed was suspended over a 100-foot cliff at an odd angle. The chains were the only things preventing the bucket loader from going over the cliff. Somebody had certainly done a good job chaining down that equipment.

The tires of the truck had dug themselves into the roadbed. The whole rig was precariously balanced. I had a winch on my Humvee but that wasn't going to help. We needed some help. My engineering battalion didn't have any large vehicle recovery equipment. I radioed the Infantry Company that we were supporting to see if they knew of a unit within the 1st ID who could help with a recovery operation.

I reached the 1st ID motor pool. They said they could send out an M78. The M78 arrived within the hour. The M78 is nicknamed the "cherry picker". It is a lightly armored vehicle with a sizeable boom extending from the back sort of like an over sized tow truck. It is generally used to recover smaller armored vehicles like Armored Personnel Carriers (APC's).

We attached the M78 cable to the flat bed and tried pulling it back up on to the road. The M78 lost traction and starting getting dragged towards the flat bed. At that point it looked to me like the M78 might just destabilize the truck and drop it off the cliff, dragging the M78 with it. I ordered a halt to the recovery operation.

The 1st ID Sgt. and his crew were starting to get excited about this recovery mission. It presented an unusual challenge. They thought we might be able to wrap the cable around some trees to give them additional support. This time I put a damper on the enthusiasm.

I didn't have much confidence in the tree idea. We would probably just end up with a tree on top of everything at the bottom of the cliff and have to pay environmental damages to South Korea in addition to road repair charges. While we were working on formulating a plan my company commander arrived in his Humvee. This was my first field exercise, having only been in Korea a few weeks. I was sure he couldn't consider this my fault. After all the equipment did belong to the company headquarters platoon, and my guys were technically tasked out. Captain Johnson took a look at the situation. The commander put me in charge of the recovery operation and drove back down the mountain. I knew then that he had complete faith in my ability to accomplish the mission. Either that, or he would rather have my name on the article 15 going to the Battalion Commander if we dropped a truck off of a cliff. Rain clouds were on the horizon. The dirt mountain road was about to become a stream.

Even if we could secure the M78, I didn't think the winch had enough power to do the job. I considered placing the M78 in a ditch on the other side of the road and trying at a different angle, but I didn't want to risk it just yet. I called the motor pool to see what else they had. It just so happened they did have one M88 in the division. That was what I needed. The motor pool sergeant agreed to send me the M88 after talking with his people on site, but he would have to round up a crew since it was evening now and everyone had been dismissed for the day.

The M88 is the largest recovery vehicle in the field. It is used for recovering 70-ton tanks. The M88 arrived after a couple hours. Fortunately the crew was in the barracks. We decided to use a plan similar to what we were going to try with the M78. We backed the M88 into the ditch. The M88 has a narrow bulldozer blade on the front of the tank chassis, so we lowered the blade to brace the M88 against the roadway well away from the trees. Gradually, we hauled the flat bed and bucket loader back on to the road. Surprisingly, the truck and trailer were fine. We unloaded the bucket loader to lighten the load, and drove the vehicles separately and slowly back to the company staging area. My commander was glad that everything was ok.

Between field exercises there wasn't very much to do in Korea, besides visiting Seoul. My interest in philosophy did not diminish while I served on active duty. I began an interesting course of study before I left Camp Mercer. I read a book about memory in the Camp Mercer library. My goal was to greatly increase my rate of learning. I had a long list of things I wanted to accomplish before I turned 30. Rapid learning would speed up all of the things I wanted to accomplish.

I found a book on memory that demonstrated the potentials of the mind. I learned that through visualization it was possible to remember and gain quick access to enormous amounts of information. I learned how to commit the order of an entire deck of cards to memory, and tell people what card they picked from where it came out of the deck.

The trick was to associate a picture with each suit plus pictures that would always represent a certain number. To remember the cards in order I would imagine a little story that linked the different pictures representing each card. This is slower than people who have truly photographic memory, but it does work. Truly photographic memory, and instant recall of information without using memory "tricks", is also possible.

I realized that the storage potential and recall abilities of the mind were unlimited for all practical purposes. I was convinced that a purely biological explanation of life could not account for people having this ability. It takes a lot of data storage to hold all the images of a lifetime plus everything one decides to imagine. Nothing in biology or cell structure accounts for that. This simple observation indicates that awareness and memory are completely independent of the material aspects of a body that can be seen under a microscope. DNA stores information to replicate cells. There is no physical explanation for the storage of memory or imagination.

When I stumble on a mystery I tend to pursue a solution wherever that search may lead, and for as long as it takes to find the solution. I finished my tour in Korea without further incident.

On the plane ride back from Korea I sat next to a Buddhist monk. Buddhism has been around for over 5,000 years. Buddhist monks of the past are credited with great accomplishments.

I talked with the monk about the mind and spirit. Buddhists believe that out of body experiences are possible and that a person continues to exist after death. Reincarnation is a strong Buddhist belief. In fact, Buddhists look for certain spiritual leaders to be reincarnated. According to Buddhism the mind, personality, and memory are all things of the spirit, which are independent from the physical body. This would explain the lack of a biological storage space for memory.

I did not consider these ideas to be that far from Christianity. The idea of Christ returning would have to be called reincarnation, no matter how people wanted to explain it. The idea of a separation between the mind and body is no different than the concept of a soul or spirit in Christianity. The fact the billions of people for thousands of years have all considered that the spirit exists, and is qualitatively different than the physical body, is a proof in itself. If something had no basis whatsoever in reality it is not likely that many person would imagine such a thing to be real, much less billions of people throughout history. I talked with the monk about my own religious background.

When I was in Junior High I joined an ecumenical church group, which encouraged attending services at different Christian denominations. I was raised a Roman Catholic. Reading through the history of the Roman Catholic Church I recognized that some Roman Catholic leaders had strayed from the ideals of Christ, the Inquisition being one example of that, so I could understand justified criticism.

I mentioned that the Roman Catholic Church had made great strides forward with Vatican II and the ecumenical movement. The idea of the ecumenical movement is to recognize similarities between Christian denominations and encourage cooperation. It is not about making everyone subservient to the Pope or changing the organization of any denomination. The Pope today does respect people from all different religions and walks of life.

The monk wondered why the Roman Catholic Church believed in celibacy, and other Christian churches did not. During the first thousand years of the Roman Catholic Church celibacy was an ideal for priests, not an enforced rule. Maybe the reason why Christ did not get married was because he had an important mission to accomplish that required all of his time. Raising a family takes a lot of time. I believe Christ lived so that everyone might enjoy family now and into the very distant future. There is a greater necessity for one person to spend all of their time on something when they are the only one doing the project. Christ was building a church that after 2000 years spans the earth today. Granted, Christ had 12 apostles. It was still a lot of work and required his complete attention. Christ never said that celibacy was a requirement. As the church grows in strength it becomes less necessary for a few to shoulder the entire burden of the many. Personally, I consider trying to enforce celibacy to be an example of ignorance regarding the nature of the life of Christ. First of all, if the greater majority of society were enlightened and doing very well physically and spiritually then there would not be much need for one person to have to spend an inordinate amount of time trying to straighten things out. The life of Christ was a celibate life out of necessity, not because there is anything good about celibacy. If everyone were celibate mankind would cease to exist. It is a slap in the face of any noble individual trying to force that individual into celibacy. If a priest decides that their help is desperately needed in a particular area, and that a personal life of celibacy is what is needed to benefit everyone else, then that should be a personal decision not an addition to the Ten Commandments created by some Pope.

I told the monk that the church was still evolving, and that maybe there would be a Vatican III and things would change. The flight back from Korea to Texas was a long flight. I spent many hours talking with the Buddhist monk about religion and philosophy.

The monk wondered how the crucifixion of Christ could have anything to do with spiritual enlightenment. What good could possibly come from being nailed to a cross? The way I saw it, the whole thing was more about a social lesson than the act of crucifixion itself. Christ did nothing but great good for everyone he came in contact with. He rarely criticized others even while they condemned him to death. The lesson is that if a society will do this kind of thing to Christ, surely a society would destroy anyone more for the sake of destruction than for any reason having to do with true justice or injustice. The lesson was that it was not enough for one individual to achieve enlightenment. That enlightenment would need to spread to most people, and a new society would have to be created before it would be possible to enjoy a better life. By the example of that sacrifice the world is a better place today. Democracy was unheard of 2000 years ago. Today, the structure is in place for a society governed by an enlightened majority, and Christianity does have the numbers to make it so.

After a long flight I arrived at Fort Hood I was assigned to a modern combat engineer platoon. My equipment included 4 Armored Personnel Carries (APC's), 4 Armored Vehicle Launched Bridges (AVLB's), 2 Combat Engineer Vehicles (CEV's), and 4 D7 bulldozers with flat bed trucks to pull the dozers. I was promoted to 1st lieutenant and served with Charlie Company of the 8th Engineer Battalion.

An APC is an armored rectangular box designed to carry a squad of 12 people. The APC has a .50 caliber machinegun, runs on treads, and is capable of deploying a smoke screen.

The AVLB is a tank with a large folded aluminum bridge on top of it. The bridge is used to cross tank-ditches or streams. Fully extended, the bridge is about 80 feet long. The bridge deploys using large hydraulic cylinders driven by the tank engine. The bridge unfolds as it deploys, setting the whole length straight up in the air before it is dropped across the gap. The bridge itself weighs half as much as the tank. While it is being lowered the back end of the tank raises up off the ground. Once the bridge is in place, hydraulic pins release it from the tank and forces are allowed to cross.

A CEV is built on an M60 tank chassis with a bulldozer blade on the front. The tank turret houses a large caliber stubby demolition gun. The demolition gun is used to lob large explosive projectiles at stationary targets. The purpose of the demo gun is to reduce barricades, bunkers, or reinforced walls. The front blade is then used to plow debris out of the way breaching the obstacle. When the

CEV is fired you can actually see the large round tumbling through the air. The CEV was designed to deliver a maximum explosive charge, not for shooting through mail slots or preventing collateral damage.

During my second year at Fort Hood my battalion was rotating out for desert training at the National Training Center (NTC). The NTC provides all the equipment and vehicles that are needed for units doing training. The NTC Red Forces were all equipped with foreign tanks and equipment to better simulate combat. Units can train at the NTC without having to convoy hundreds of APCs halfway across the country.

That means that all the unit's equipment left behind in the motor pool still needs to be maintained. My platoon had the most equipment in the battalion since it was a specialty platoon structured for bridging operations and breaching obstacles. A typical combat engineer platoon consists of 4 APCs. Keeping the equipment operational was a command priority. My battalion commander was rated on equipment readiness among other things.

Someone had to be left behind to act as rear detachment commander and oversee maintenance of the battalion's equipment, daily PT, and whatever else might be required. That turned out to be me.

You might have heard about the Branch Davidian standoff in Waco, Texas. Waco is located about 80 miles Northeast of Fort Hood. The Branch Davidians received broad news coverage. One couldn't help following it on TV or looking at a newspaper during March of 1993.

David Koresh was the leader of the Branch Davidians. He was considered to be a prophet by his followers. The Branch Davidians believed that they would play a role in preparing for the Second Coming of Christ. David Koresh focused on interpreting Revelation. He believed an Apocalypse would occur in the near future so he built a survival facility called Mount Carmel in Waco. At Mount Carmel David Koresh stockpiled food and weapons. No illegal weapons were ever discovered at Mount Carmel.

The press coverage during this standoff gave people the impression that the Branch Davidians were set up in bunkers with machine guns and rocket launchers. The original raid conducted by the Bureau of Alcohol, Tobacco, and Firearms (BATF), was for a failure to properly register legal weapons not for the possession of illegal weapons. The only charge was a lack of paper work that the Branch Davidians might have resolved without going to court. That isn't the way things worked out.

The BATF raided the Mount Carmel facility in SWAT team fashion. It is not clear who fired first. There is some speculation that BATF initially shot each other during the raid, triggering a violent exchange with the Branch Davidians who were suddenly attacked with deadly force on their own property. The end result of that initial raid was 4 dead and 24 wounded BATF agents; 6 dead and at least 1 wounded Branch Davidian. We will never know exactly what happened at Mount Carmel. By the end of the 51-day standoff all 70 of the Branch Davidians were dead including 21 children and many young women.

After the initial BATF raid the situation was turned over to the FBI. An effort was made to negotiate surrender. David Koresh made several offers to surrender with minimal requirements like printing his writings and his viewpoint of the situation in the press. No additional law enforcement casualties were reported during the siege of Mount Carmel. The FBI decided they did not want to wait any longer. They called up nearby Fort Hood requesting tank support for a final assault on the compound. That assault would result in no survivors.

At that time I was still rear detachment commander for the 8th Engineer Battalion. The Brigade Commander I reported to asked me to prepare a CEV and crew to support the FBI in Waco. There were only two engineer battalions stationed at Fort Hood with CEVs. There was a 50 / 50 chance that a CEV under my command would be needed.

I gave this some serious thought. The U.S. military is commissioned, trained, and equipped to handle foreign military aggression not civilian law enforcement. Civilian law enforcement is an entirely different mission than the mission of U.S. armed forces. The idea of law enforcement is to use the minimum force required to safely apprehend individuals and bring them to justice for a fair trial by jury.

That is not the mission that most military forces are trained for. The military is trained to maximize firepower on a particular target, and win quickly with minimum friendly casualties. Some special forces are trained for urban warfare and do practice clearing buildings with non-combatants present. All military officers are trained in basic rules of war. The U.S. military does care about people, and does try to protect innocent women and children.

At the end of the day, the CEV sent to the FBI to support the assault on the Branch Davidian compound came from the other engineering battalion stationed at Fort Hood. Having been asked to keep my equipment at the ready was nothing more than the job I was commissioned to do. Having presented my reasoning to our brigade commander, he realized I would question a direct order to deploy tanks against U.S. citizens who intended to surrender.

The standoff could have, and should have, ended peacefully. Setting the precedent of using the U.S. military against U.S. citizens is a dangerous precedent. It shouldn't even be considered in anything but the most desperate of situations, and even then it should require the approval of the President of the United States. That is my opinion. If a situation arose that involved a radical group of U.S. citizens marching on a city with weapons of mass destruction in hand, certainly the use of military force would be justified.

The FBI claims that tear gas was used in the final assault to speed up surrender, and that the tanks were simply used to approach firing positions safely to fire tear gas into the structure. The U.S. department of defense has a wide array of munitions to choose from. Tear gas is certainly one of them. Tear gas can be deployed in hand held canisters, or fired from weapons of various calibers. Some tear gas rounds are secondary incendiary devices. In other words, they can start fires. Tear gas can be lethal in a closed room. The military trains with tear gas. I doubt the FBI is ever exposed to tear gas to learn exactly how debilitating it can be.

Even out doors, if people don't escape the tear gas area quickly they will be temporarily blinded and usually on the ground throwing up. Having been exposed to tear gas, I can vouch for that. The purpose of that training is to instill the importance of putting on a gas mask quickly, and that training is not easily forgotten. I have seen tear gas canisters start fires out doors, never mind inside a wooden structure. It doesn't take much thinking to realize that people would be immediately blinded and unable to escape after a tear gas assault on the Branch Davidian compound. A room full of tear gas immediately deprives a body of oxygen. I would guess the people didn't have time to pull in and hold a breath of fresh air, close their eyes, and run quick enough to the door by memory. After a whiff of tear gas, I can hardly imagine a lung full of it. Tear gas makes the eyes feel like they are burning, so I imagine it would be the same inside the lungs. In any case, combine that with the rapidly spreading fire caused by the assault, and it is not hard to see why no one survived.

It was easy for the FBI to claim that the Branch Davidians started the fire when there were no surviving Branch Davidians to claim otherwise. This incident brings up an important point. If government agencies will do these things in relatively calm times, and if the general public doesn't care enough to say much about it, what do you think will happen during difficult times? This nation is closer than people might want to imagine from a draconian police state run by an elite group who care nothing about the general population beyond wondering if the military can be used against that population.

I know that sounds extreme. Let us look at both sides of the argument. An elite group with plenty of funding could argue that the general public is not particularly bright, and unwilling to work together politically to accomplish a massive survival project that would be needed to safe guard an entire population. People who are concerned mostly about themselves would argue that it is best not to tell the public, make sure the general public can be controlled, and focus resources on certain very small groups. I do not believe that I am the only one who understands the technical certainty of the events this planet will face in the near term.

Up until now, special interest groups have accounted for most political campaign support. It would be easy for a few people to financially back sincere people and pull strings as necessary within the ethical limits of the person put into office. It is easy to justify the actions at Mount Carmel if people believe it was simply a mistake. If that action was a test to determine the willingness of the military to support action against civilians that did not pose an immediate threat, that would be a different story. The extreme circumstances of the incident warrant further investigation.

If an operation requires military equipment then the mission should be turned over to a military officer who is familiar with military equipment and rules of war. One could do no worse than a 100 percent enemy casualty rate and no prisoners taken during an assault. I do not believe David Koresh set fire to his own children after his room and every other room in that building filled with tear gas.

A conventional assault would have involved shooting anyone actively resisting, then accepting the surrender of the remainder including the women and children. The FBI acted in a cowardly manner, lacking confidence in their ability to take a building conventionally. The FBI was unwilling to take even the slightest personal risk. The FBI was unwilling to be inconvenienced by waiting out the Branch Davidians. If the FBI had bulldozed down the walls of the place with the CEV blade, and removed the Branch Davidian supplies, that would have been a quick resolution. There would have been no risk to the tank driver. Granted, hind site is 20/20. Even so, a long list of less fatal options did exist.

I continued serving as a platoon leader in the 8th Engineer battalion over the next year. During that time I decided to change my name. I found out that changing one's name involved getting a lawyer, going to court, and getting a clearance from the FBI. The red tape didn't bother me, and I went through with it. My given name is Auger. Back in the formative years of this country many people changed their name after arriving in America. That is why there are so many people going by the name of Smith. It was not uncommon for people to call the town smith, Smith.

In the English language an auger is a wood working tool. Way back in the family tree a relative of mine changed the family name after arriving in America.

Anyway, I wanted to select a name that fit what I wanted to do with my life. Paladin has a couple of definitions. The first is a knight who champions a royal court. The second definition is an outstanding protagonist of a cause. That fit so I changed my name to Paladin. Later I found out there was a TV show about a guy who traveled around the country setting things right. He passed out a business card that said "have gun, will travel!"

I spent another 6 months at Fort Hood completing my four-year active duty commitment. I decided to transfer to the reserves after completing my tour on active duty. A couple programs exist to place military officers in management positions, usually at fortune 500 companies around the country. I thought about enrolling in one of those programs. That would have required being flexible about where I wanted to live next though, and I had my heart set on Austin, Texas.

A few different things drew me to Austin. Austin is the closest major city to Fort Hood, so I had spent some time visiting Austin on the weekends. Austin nightlife is centered on 6th Street. The street is lined with clubs, and about half of them have live entertainment. I traveled to Austin often while at was stationed at Fort Hood.

One weekend I was touring around Austin by the University of Texas looking at taking classes towards an MBA. I noticed a Church of Scientology across from the University. I knew very little about Scientology. The concept of integrating science and religion sounded like an unlimited vista for philosophical debate. I decided to learn more about it, and I started a course at the Church of Scientology in Texas. I also accepted an engineering position at MLC Cad Systems in Austin, Texas.

Scientology and Buddhism

Learning more about Scientology is not a weekend affair. The founder of Scientology, L Ron Hubbard (LRH), recorded over 1000 lectures and published well over 100 books. I enrolled in a basic course and began studying Scientology. I was in the middle of one such course while transferring from active duty to the reserves. That was one reason I decided to move to Austin besides just really liking the city. I studied Scientology in great detail.

There is a much truth in Scientology. More than Buddhism, which is a huge subject itself. At the same time, some basic fundamentals of creation and life were missed by LRH, or poorly understood. After the death of LRH, much like after the death of Einstein, original research stopped in Scientology.

I am sure that LRH would have be quite disappointed by this were he around to say anything, but he is not, and the existing leaders of Scientology think that there is nothing more to be discovered beyond what LRH already wrote down. So, I do not expect to see any further progress on the path of knowledge within the existing Scientology organization.

Knowledge is a pyramid. LRH came very close to the top of that pyramid. In fact, judging by his work presented in a lecture series called "The perception of truth", LRH actually glimpsed the top of the pyramid. He abandoned that line of reasoning though, maybe because he would have been forced to accept the existence of a universal creator if he followed through to the logical conclusion. Instead, LRH focused on a narrow viewpoint involving improvement at an individual level. Things like solving grand unification, figuring out what is happening with this planet, how the universe was created, God, angelic spirits, and other big questions were not the focus of LRH's work.

If LRH did understand the physics of this planet he did not write about it publicly. The man did say the planet was facing a near term catastrophic challenge, but he said nothing regarding the details or reason why. He could just as easily have been concerned about the possibility of nuclear war as anything else. Nothing can be done about a vague threat. Precise intelligence is needed to take action.

Scientologists believe in reincarnation just as Buddhists do. I know that reincarnation is possible, and that anyone has the potential of being reincarnated. I also believe that the individual spirit has the potential of not being reincarnated. In other words, a soul might rise to greater heights closer to God, or a soul might fall to lower lows and never walk this earth again nor ever see it from on high. A variety of possibilities exist for the individual spirit.

LRH recognized that electrical engineers had missed the philosophical importance of a ground, or base, in electronic circuits. Today's electronics need to be grounded to work. That ground is tied to the planet, which is anchored in the solar system, which is anchored in the galaxy, etc. LRH believed that life itself could create energy within this universe and do other neat things. This is certainly true. The big snag LRH ran into was when he asked himself the question "What is the source of life?" This question drove him mad to some degree, and he decided to stop thinking about that. He also abandoned the search for a universal electronic base.

One can assume that the spirit of man is the greatest thing in this universe. One can assume anything. Sometimes even a wrong assumption helps to accomplish certain projects. Looking at big projects like predicting the future of a planet, or working directly with the energy of creation, the right answers are needed to get results.

I studied Scientology part time for about 5 years while I lived in Austin. The main reason why I moved to Austin was because I was offered a position at MLC Cad Systems working with high-end 3D solid modeling systems.

Engineering

My job at MLC Cad Systems involved visiting many different technology companies. I brushed up on my design skills and learned about the latest technologies in industry. This was good since I had been out of circulation for four years on active duty. I also gained some sales experience.

After two years at MLC Cad Systems I was offered a position at Scientific Measurement Systems (SMS) as their manager of scanning services. SMS was a very interesting company doing 3 dimensional X-ray scanning. I wanted to work with that technology and gain additional management experience so I applied for the position. SMS offered me a good stock option plan plus a good salary. Those stock options turned out to be a very good deal two years down the road.

3D X-ray scanning is like an industrial CAT scan. Both the insides and outside of objects can be viewed and measured without opening the object. The only difference between a CAT scan and industrial CT scan is the energy level. A much higher energy X-ray beam is needed to penetrate an engine or a whole car for example. Most of the jobs we did came from the automotive industry and defense. We did other projects as well, like scanning a Barbie Doll. It is difficult to scan a Barbie Doll with traditional laser scanning. Areas on the torso are difficult to reach in a direct line being blocked by an arm.

The CT scanner is a good analysis tool. We did projects on a regular basis for a company called Failure Analysis. Sometimes a court case would require detailed information about the parts that failed. We could scan a whole casting for example and determine what the level of porosity was in the casting, which can reduce the strength of a part.

You would be surprised what objects people want to X-ray scan. We did one project valued at $40,000 scanning diaper samples for a major diaper manufacturer. These samples were from competitors. Apparently the competition really was more absorbent and they wanted to know why.

We scanned new shell casing designs for M1 tank rounds, and even a number of space shuttle components. Space shuttle components are very expensive. We scanned a small rocket engine used to adjust the attitude of the space shuttle main engines. These thrusters used to orient the main engines had a history of failure in space. The shuttle can still navigate with only two out of three of these small thrusters. People get worried when one goes out.

During one shuttle mission there was a problem with one of these small thrusters. We were asked to do some high-resolution scans on one little rocket engine that had been commissioned to fly on the shuttle. The idea was to try and learn more about how wear was affecting these engines to understand the cause of failure. It was an interesting project. We did scanning 24 hours a day for nearly a week. We did find some small pits inside the combustion chamber that may have been significant. Once a component is approved to fly it cannot be taken apart without going through the whole validation process again, which is an expensive process.

After working for SMS for a couple years I decided to pursue some different projects. It was 1997, during the boom days of the market, and my SMS shares increased 10 fold. A rising tide takes all ships with it.

I decided to sell while the share value was high. This was the first time I had any significant amount of money in one place. I decided to spend it. Actually, I did have a plan that involved investing it in a new business.

Health Products

I have always been interested in health and fitness. Someone sent me an audio tape about colloidal minerals called "Dead doctors don't lie". The tape was convincing. It looked like people were getting rich doing multi level marketing by just mailing out a copy of this tape to mailing lists. I spent $8,000 on tapes, mailers, etc. and mailed out something like 1500 copies. I took about 150 orders, signing up that many new members.

The tape I mailed talked about people who lived very long healthy lives living in high altitude locations near glaciers. The people drank water from glacier run off, and the idea was that the minerals in the glacier water benefited people health. That assumption is not necessarily true. The water itself running off from the glacier might have been qualitatively different. For example, some water is acid water, other water is neutral, and some people swear by machines that produce alkaline water. There is also something called micro water that has smaller clusters of water molecules within the water.

There are only a couple colloidal minerals that I know work very well. The colloidal mineral product I was selling had too many different kinds of minerals, many of which were not helpful, and very low amounts of the ones that are helpful. I was also selling vitamins but the cost of the products in this multi level marketing structure were too high compared to equivalent products at retail outlets.

Here is the deal on colloidal minerals. Prior to the FDA being formed in 1938, silver dominated the medical field as an antibiotic. When the patents ran out on colloidal silver, it was not profitable for the medical industry to sell it, so it went out of favor. The product was talked down; citing probably one case in a million where someone's skin took on a bluish tint because they ingested very large quantities of chemically produced silver products.

The truth is that most strains of bacteria, even some viruses, are killed immediately by colloidal silver. Many strains that have become resistant to today's antibiotics can be handled by colloidal silver. It is not possible for the bacteria to mutate against silver.

I now produce my own colloidal silver using a low voltage technique resulting in very small particle sizes. This process takes some time to complete. A batch takes 10 hours to run compared to some machines that process a batch in 20 minutes. The silver kills single celled organisms by shutting down their anaerobic respiration system. A virus is much smaller than a bacterium so it stands to reason that the smaller particle size would do better against viruses, which are more difficult to handle than bacteria.

There has been some progress with companies doing double blind and triple blind studies trying to make some impact on the FDA. There is no reason to prove anything to anybody. Testimonials, and personal experience, speak for benefits of colloidal silver.

Colloidal gold is the second colloidal mineral that I have found to be beneficial. It helps the nervous system and improves reaction times by improving electrical flows in the body. There is more to the body than meets the eye. Even CAT scan technology can detect a certain "aura" around a body. A little colloidal gold might be improving energy to bring about an increased sense of well being. However the colloidal gold works, I know that it does work. I have seen reports indicating that gold improves the connections between nervous system cells.

A little colloidal silver and gold are helpful. I use special water to make my own colloidal silver and gold. The water is key, and what I use is at least as good as the glacial run off.

Very few of those people I signed up brought other people into New Vision. After 3 or 4 months I began to miss that regular paycheck. I didn't notice any major change or improved energy level drinking the New Vision colloidal minerals. I ended up going back to work fulltime.

The Oil Industry

After trying to become a multi level marketing millionaire, I went back to engineering work. I worked for a couple different companies over the course of about a year before taking a position at Varco International. At Varco I did design work on a variety of oil industry related products.

The oil industry is a fast moving business. There are a number of reasons why that is. The biggest reason is that drilling for oil requires a lot of money for both equipment and personnel. Large drilling rigs, and mobile drilling platforms, spend a certain amount of time on-site drilling the oil wells. When that is done they move on to a new project.

Since certain equipment is needed for drilling the initial well, and different equipment is needed to pump oil out of the well if oil is found, capital equipment is usually sold by day rates rather than purchased outright. Granted, some larger companies own all the equipment needed to get a well into production. They still want to make the most efficient use of that expensive equipment.

I spent a couple weeks working off the coast of France on a semi-submersible platform called Energy. The platform was in the final stages of construction. The Elf Corporation based in France was anxious to get that platform moving out to a drill site somewhere in the Atlantic Ocean. Varco International was responsible for the drilling instrumentation and control package. The system looked sort of like a Star Ship Enterprise captain's chair. In front of the chair we had an array of touch screens and joystick controls. We ended up working close to 18 hours a day finalizing that installation.

A drilling platform out on the ocean is an isolated little community. The Vikings many years ago would have been in awe seeing these huge platforms and large ships like aircraft carriers. I was impressed. Before flying back to the U.S. I had a couple free days to explore Brest, on the coast of France.

They still call clubs Discos in France. I was traveling with another design engineer named Ron Exley. Ron never left home without his cowboy hat and boots, France or no France. Ron had a pub staked out with pool tables and cold beer. They serve just as much warm beer in France as they do cold beer, so it is important to do advanced reconnaissance (recon).

I did some karaoke at the pub, and met some girls there. I told them I was from Texas, which seemed to be a popular state over in France. They brought me to one of the discos later. It was an enjoyable evening. Fortunately, Ron managed to get us to the airport in time for the flight the next day.

Music

Shortly after I started working for Varco my part time focus shifted to learning about music. My goal was, and still is, to perform on stage. I started practicing keyboards during the evenings and weekends using both a Korg Trinity and a Korg Z1 that I had purchased.

After spending a great deal of time and effort learning to play only a couple songs, I realized that learning to play keyboards was not an easy task. At that rate, learning to play a minimum of a few dozens songs that I would need to do a show was going to take a very long time. My real interest was singing and song writing, but I didn't want to give up on playing music.

I thought long and hard about this. I looked at what exactly was slowing down my progress learning to play these songs. Being able to move fast enough to the next note or chord was certainly one thing. I decided to write a real time software application to give me a performance edge.

The idea was that the software would correct any errors I made in real time so that the error would not mess up the performance. For example, if I hit three notes for a chord, and the third key I hit was not the right note for that chord, the software would correct that note. The software would take the velocity I hit the third key with, and sound the proper note at the proper velocity, rather than sounding the wrong note. When I released the wrong note, the software would release the right note.

Sounds complex. It was a very complex program to write. Fortunately, it did make playing much easier. This whole system used a MIDI keyboard and sent the processed output to a midi output device. The output device could be a software synthesizer or the sound generator on the Korg keyboard that I was playing. The software worked with a MIDI file of the song I would be performing.

After getting that working, I decided to expand the whole project and build an accompaniment system which allowed other tracks in the MIDI file to follow along precisely with the computer assisted track that I was playing. I called the application Maestro.

Maestro was a major development project. Since I didn't make millions in the multi level marketing business I figured I would make millions in the music software business. I would use that income to finance a band and break into the music industry.

I sold a few copies of Maestro on line. I didn't sell enough to make a living at it though. I decided to use some vacation time to attend the National Association of Music Merchants (NAMM) trade show out in LA.

The winter NAMM show is a huge music trade show that fills up the entire LA Convention Center and then some. The trade show itself is a four-day event. Buyers from music retail stores, and people involved with music publications, walk around the show looking at new music related products. My current release of Maestro had not been thoroughly beta tested yet. I didn't have a plan to do product support or for following up on leads. I was still doing engineering work full time. The way I looked at it, at the very least I would gain some experience attending the trade show.

My brother Marcus is an accomplished singer and guitar player. He graduated from Syracuse University with a degree in vocal performance. Marcus was doing some auditions around the country to make a career of singing, specifically opera singing. He had also recently started a mortgage business with a couple friends who were partners. They were doing very well. I asked Marcus to go out to LA with me to help run the booth at the NAMM show. He said that he would.

I arrived in LA the day before Marcus did. I checked out the convention center, and the next day I picked Marcus up at the airport. I had a hotel room within walking distance of the convention center, which made things convenient.

We set up all the equipment the following day and started the show. We had a Korg keyboard, speakers, and a projection system to show the Maestro software interface. Marcus took to my new software quite well. He started playing songs standing behind the keyboard with his eyes closed. Naturally, people thought he was a very talented Sammy Davis junior able to play upside down.

Not knowing he had any help from a real time error correction software system, people were quite impressed. It was a good show. We met some good people. There was an engineer who stopped by from Disney World who was working on a new attraction to give people the experience of playing in a band without having to spend years learning an instrument. He had a team of people working on the project. His approach was a little different than mine. The main difference was that they were using digital audio and I was doing things with MIDI. He promised to pick up a copy of the software after the show and stay in touch.

During the first few days we sold a few copies of software and talked with a few interested retailers. Had I really been ready to ship a product I am sure I could have taken some good orders. I wasn't ready for that yet though, so we just collected business cards.

On the last day of the show an eccentric character stopped by the Maestro booth named Bill Glasser. Bill is great guy. Bill played around with the software and was excited about the product's potential. He thought that he could help us raise money for the project and asked if my brother and I wanted to go for a beer after the show.

Since this was our first trade showing we were located in the smallest booth available in a back corner near the fire escape. It was probably obvious that we were not very well funded. We exchanged business cards and cell phone numbers. Bill had a restaurant in mind that served great food and had good German beer.

After the show we managed to link up with Bill in the crowd and then made our way to dinner. Bill's general idea was that I could sell a controlling interest in the project to investors. I would serve as CEO and manage the company. This would give investors the security they needed to make a significant contribution to the company.

I was very possessive regarding my software project. I didn't want to give up control of it. That is what I told Bill, and he accepted that. We agreed to stay in touch. Looking back at that, the structure that Bill was proposing was fair.

The night before leaving LA my brother and I decided to check out sunset strip in LA. We ended up at an Irish tavern with pool tables. We were having fun. I thought of an idea to meet some other people at the tavern. I dared Marcus to go sing an aria (solo vocal performance) to one of the girls sitting on a couch inside the tavern. Marcus is not shy. He knows quite a few pieces by memory. Marcus walked over and asked the girl what songs she liked. She knew some opera songs, and selected one that Marcus knew. Marcus sang the whole song at the top of his lungs, which is pretty loud, and drowned out the club speakers without a microphone. He won the dare, and I bought the next round.

Not long after I returned to Austin, and Marcus arrived back in Georgia, life started to get interesting. Not along the lines I expected though. I did some follow-up after the trade show but didn't actually close any distribution deals. Three months after the show Marcus called up to tell me about a commercial loan deal he was working on. This deal involved raising money to retrofit an old power plant with a new energy technology. The name of the company was Laser Fusion Technologies (LFT).

2

Fusion

Due to the sensitive nature of this chapter and the next, character names, companies, and technical descriptions have been fictionalized as necessary. LFT is an ongoing project that probably represents the most important energy breakthrough being used by the Department of Defense (DOD) today.

The sensitive nature of this project is not related to patents. The most important technologies are not patented. It would be naïve to assume that elected officials, including the president, know everything about what is going on in the DOD. Elected officials, including the president, serve limited terms and are not always kept informed. Every president elected in the past has been influenced if not selected by major party bosses. That truth goes very deep. Only one president in the past 200 years who actually won the presidency after deciding to run without prior approval, or without being invited by, the leaders of a major party. Care to guess who that was? That president was John F. Kennedy. Kennedy was also one of the few presidents to be assassinated. When a single private individual decides who will be funded, and who will represent the party, that one individual wields great power. One could question the motivations of certain party bosses. Personally, I do like the Republican Party, and I am not trying to blame anyone or suggest a conspiracy of any sort. I am just pointing out that presidents have been selected to serve particular purposes, and that those purposes are not always known or in the best interests of the majority. An individual can only act on personal knowledge. President Bush is a very sincere individual who is trying to act in the best interests of the nation. At the same time, President Bush was an average student, without a technical background, who would not even try to get to the bottom of details regarding technology.

National secrets of a technological nature do exist. A significant number of patents each year are stamped "National Secret" and returned with a note listing penalties for divulging the secret stumbled upon by an ambitious inventor. Those penalties include long jail terms, or worse, should the inventor talk to anyone about the invention, much less try to build it or market it. That is the reality of the situation. Considering some inventions like building a nuclear bomb could have dire consequences, it is no surprise. The executive branch is responsible for national secrecy policy. The president is never presented with the details of technology related to national secrecy, that information is given on a need to know basis to create certain defense related products. The important point is that the president is the only person with the legal authority to change existing national secrecy policy.

Here is the national security analysis regarding fusion. Fusion represents an enormous amount of energy in a small package. The only practical use anyone has made of fusion publicly has been testing hydrogen bombs. Hydrogen bombs are not very practical from a military point of view, much less from the viewpoint of all mankind. On the other hand, a controlled fusion reaction produces very little if any harmful radiation. Controlled fusion would be the ultimate conventional weapon system. The kinds of weapons that could be produced would be similar to what we see in the movies. For example, compressing a very high-energy industrial laser into a hand carried device would give you something like a Star Trek phaser.

Controlled fusion is just the tip of the iceberg regarding new weapon systems.

A fusion breakthrough could be used on a field of battle to defeat conventional U.S. weapon systems like the M1 tank or tactical aircraft. If such a breakthrough were made in the U.S. a decision would have to be made about what to do with the technology. Filing a patent is no different than going public. A patent makes a design available to everyone who wants to look at it, including people in other countries. That is why it is illegal to patent a nuclear bomb.

Now lets look at a fusion breakthrough from an economic and environmental point of view. An economical fusion reactor would greatly reduce the cost of energy. This would result in rapid economic expansion. Since fusion is a clean renewable energy source we would not have to worry about running out of oil. Obviously this is a great benefit. Does it outweigh the risk of publicly releasing the technology? Up until 2003 the choice has been to classify all workable fusion designs. Oddly enough, I believe that has happened without people really understanding why these designs work in the first place.

Here is one more thing to consider about a fusion breakthrough. If a workable fusion design validated a new physics model that had significant bearing on the future of mankind, should that fusion breakthrough be made public? If the future of mankind were at stake I would hope that this would sway the opinion of a few people managing black budget projects.

Before I start talking about LFT, let us consider how one goes about solving difficult technical projects like fusion. There are two ways to tackle any tough technical problem. Those two logical paths are deductive and inductive reasoning. Deductive reasoning goes from the specific to the general. Trial and error, or accidental discovery, brings about the use of deductive reasoning.

For example, a researcher makes a startling new discovery. A compass needle moves off magnetic North in the vicinity of a charged electric wire! This is a discovery that actually occurred. As a result, many years later, all the laws were written concerning relationships between electricity and magnetism. These laws are called Maxwell's laws, because Maxwell wrote the laws after studying electricity and magnetism. People notice a phenomena, do a whole bunch of experiments, and figured out how things generally work with electricity and magnetism. This is called deductive reasoning. You can see that the general rule discovered by deductive reasoning is not always ultimate truth. The rules are only true based on the particular phenomena that were originally noticed. Maxwell did not study or even consider gravity when he wrote Maxwell's laws.

Inductive reasoning works by going from the general to the specific. This is very different. The only way to start with a general law is to just invent the law or have an epiphany. Once a general law is conceived it can be called a hypothesis. Here is a hypothesis. Fusion is the result of particles entering a strong gravitational field that exists at the center of a large spherical energy field. Using this hypothesis, an engineer would then try to figure out ways to create a large spherical energy structure to test the hypothesis.

Trial and error is part of deductive reasoning. By trial and error one could throw a bunch of things together, mixed with some inspiration, and accidentally producing a stable spherical energy structure. The trial and error researcher might then observe a large energy release and a little helium being produced in the reactor. Does that mean the trial and error researcher knows what caused the fusion energy release? It does not. If the reactor chamber was very complex, and difficult to see directly, the researcher might never know exactly how or why fusion was happening in the reactor. The hypothesis I mentioned might never be considered, and the company might continue for years producing reactors that worked but not knowing why.

A team could go down this trial and error path never realizing they had discovered or created something that explained many of the mysteries in physics today. Never mind making any connection between their work and the current state and future of our planet. With that in mind, let's continue the LFT story.

Marcus began working with LFT at a very early stage in the development of the company. LFT was created to do waste management research. The idea was to neutralize toxic liquid wastes by burning them at extremely high temperatures within a reactor that used an electromagnetic containment field. Large industrial lasers were used to turn the waste stream into plasma that would be contained by the containment field. Converting the waste into plasma would break all the molecular bonds turning it all into basic elements rather than toxic chemical by-products.

Barry Elwood was the president of LFT. He designed LFT's three prototype reactors. Barry personally funded the initial phases of the LFT project. He also bought out another company that was involved in similar research.

Barry recruited some friends to help him get his new venture started. Frank Wellington signed on board to serve as Vice President of Operations. Frank had experience managing engineers, and general good business sense. Much of the mechanical engineering work and fabrication of components was being contracted out at first.

Barry completed a first generation prototype of his reactor that produced some very interesting results.

Frank's initial focus would be planning and preparing for due diligence visits being conducted by a number of major corporations and investors. LFT had a prototype that would impress people. Bob Wainman, a friend of Barry's, put in the first $1 million to build a larger second generation prototype. Bob served as a key member of the board for LFT.

LFT leased some industrial space on the outskirts of a small town where they would build systems and host visits.

Marcus got involved with LFT through Tom Bearden. Tom was a financial consultant specializing in raising money for early stage ventures. Tom had been working with LFT ever since Barry started raising money to expand the project.

Marcus was doing well with his mortgage business. He invested mostly in the public stock market. To round out his portfolio Marcus added a couple private venture positions. One of these ventures involved technology used in photography. The second venture was a company called Immunex, which had demonstrated a vaccine for AIDS over in Africa. Tom recommended both of these private ventures.

Tom knew that Marcus was getting into the commercial loan business. He thought Marcus could help him with funding for LFT. Tom gave Marcus a call and arranged a meeting to discuss LFT.

The latest LFT reactor was a 10-kilowatt device that achieved a 20 X energy gain. This gain was calculated by comparing the chemical energy contained in the reactor fuel with the energy contained in the heat that the reactor produced. Once the reaction was started it would run continuously with a fuel mixture containing a significant percentage of water.

The fuel mixture was not critical, as long as it was a mixture of carbon-based fuels and water. The reactor produced no significant radiation. The by-product was a solid material that collected at the bottom of the reactor chamber. This waste material was clean and could be used for road building or construction if someone wanted to.

LFT's reactor had an array of sensors installed to detect gases, temperatures, flow rates, electrical energy, and magnetic fields. During one test run of the reactor the large industrial lasers they were using shut down due to a minor controller failure. Without the lasers running, but with the containment field intact, the reactor continued to run with no significant drop in energy output. That means that the enormous temperatures achieved using the lasers continue to exist without the lasers. That implied that something other than a purely chemical reaction was taking place.

Test runs of the reactor were limited to an hour or two because of problems with cooling of components. Temperatures at the core of the reactor were estimated at about 1 million degrees. Small quantities of helium gas were detected in the reactor chamber indicating that a fusion process was responsible for the energy release.

LFT had conducted extensive due diligence testing with Sandia National Labs, Bellinghouse, and Mobex among others. Barry, the founder of LFT, had been working closely with Dr. Speilvogel, head of the Z-Pinch project at Sandia National labs. I am not fictionalizing known government departments, just private companies and related people.

Sandia Labs wrote letters of recommendation for LFT saying the technology was the most promising thing they had investigated so far regarding new energy sources. Sandia National Labs investigates dozens of projects, including the various cold fusion claims that were in the public eye for a while. The vast majority of these claims turn out to be false, or the result of honest mistakes in the lab due to faulty measuring equipment or inconsistent results. Some claims cannot be reproduced. LFT's reactor worked every time, up until it melted down key components, then it needed some fixing for another run. The reactor was not radioactive. Unlike fission, not all fusion reactions produce harmful radiation.

Bellinghouse was extremely impressed. So was Mobex. Barry and the rest of his management team entered into negotiations with Mobex while Tom was involved in the project. A month prior to Marcus arriving on the scene Barry decided to walk out of the Mobex negotiations. They left 50 million dollars on the table, but Barry retained control of his company.

The impression was that Mobex would be willing to go much higher in terms of putting cash into the project, however Mobex would not compromise about gaining control. Fifty million was enough to take their next step and retrofit an older power plant with this new technology. Barry and the rest of the guys at LFT realized how important this technology was to the world. They did not want to sell it out and risk it being put on the shelf of a major oil company.

The LFT financial plan was to raise 50 million dollars to build a power plant. They would use revenues from the energy produced to expand. Internationally, LFT wanted to license the technology to larger corporations that could handle deploying this technology in their part of the world.

Barry, Bob, and Frank realized the future earning potential of LFT. Selling ownership in the company at this stage would be expensive, compared to the cost of debt financing. Ownership represents a percentage of all future earnings, which LFT valued at tens of billions if not hundreds of billions of dollars. I would be better If people really understood Einstein they would have asked where the missing dark energy was in the universe, not "dark matter". to borrow 50 million dollars in the short term than to sell 40 percent ownership. That ownership could end up being worth 40 billion dollars.

Tom recommended that Frank talk to Marcus about a commercial loan. Tom had an agreement in place with LFT that provided compensations for fund raising depending on whether it was equity or debt financing. Tom was willing to share this with Marcus.

Marcus received LFT's business plan and talked it over with Tom before scheduling a visit. Frank agreed to meet Marcus at the mortgage company office. They talked about what LFT wanted to achieve both here in the U.S. and internationally.

At first Marcus was skeptical about the project having read all the articles on cold fusion. Frank scheduled a visit for Marcus to tour LFT and see the reactors. Marcus was very impressed with the huge industrial lasers, the reactor itself, and the melted down remains of earlier prototype systems.

After returning to his office Marcus did a business review for the project. LFT had no operating history and they were trying to finance a completely new technology that had not been proven. Marcus considered LFT's chances of getting a commercial loan close to zero without close to 100 percent collateral. Barry and Bob had spent a good percentage of their money getting to where they were. They were not in a position to put up collateral for 50 million dollars. As it was, the company was running on a tight budget. What they were currently doing, trying to raise money from private investors, was one of their best options.

A huge company like Mobex can put in tens of millions of dollars and barely stress their cash flow. It would not be in the best interests of Mobex to rush a technology to market that would put their existing business out of business. When the world is close to running out of that non-renewable it would make sense for Mobex to diversify. Who knows what Mobex has already bought up and put on shelves. They could already have technology in the wings that is more economical to deploy than what LFT was demonstrating. If that were the case their only motivation would be to buy out possible future competition not to gain more technology.

On the other hand, a private investor is most concerned with return on equity. Some private investors want to play a role in day-to-day operations but many do not. LFT had the attention of larger corporations. They did not have a large database of private investors. Most private investors get in around the $10,000–$50,000 range. Some do deals in the $250,000 range. Raising 50 million dollars would take quite a few private investors. Building a power plant takes money.

Marcus recalled our meeting with Bill Glasser at NAMM and thought he might be able to help bring in some private investors. It was a long shot, but it was worth a try.

Marcus had another idea as well, that involved some kind of joint venture. Frank had mentioned licensing the technology internationally. Maybe a company overseas, like Mitsubishi, would help fund the development of the technology in exchange for exclusive rights within a certain geographic area. This would allow LFT to retain control of the technology and earn royalties without having to build a whole infrastructure in a foreign country. LFT's main interest was focusing on the U.S. They wanted to build a reactor manufacturing facility locally to boost the economy and create jobs.

Marcus spent a week or two trying to develop contacts at Mitsubishi. One individual promised to bring the LFT project to a director of U.S. operations for Mitsubishi. He was sure the director would be interested in the project. He only wanted a five thousand dollar fee in advance to make the introduction.

Marcus seriously considered the offer. He ran the idea by Frank at LFT. Frank said that they had already paid certain individuals fees to help with financing and they were not impressed with the results.

Marcus gave me a call shortly after that. He brought me up to speed on the LFT project and talked about his visit to the company. He asked if I still had Bill's card and if I would try and help with the project.

When Marcus said he was working on funding a fusion project I wondered if he was being taken in by some sort of con artist game. I was extremely interested in fusion. I considered it the single most important leap that mankind needed to make if we were ever going to advance much beyond being a fire culture. I really don't consider burning oil much more advanced than cave men burning wood. Sure, we forge complicated looking engines. People have been forging metal for thousands of years.

I was skeptical for two reasons. First, all the home brewed cold fusion methods had not worked. Second, I was surprised that my little brother would be working on funding the single most important project in the nation, if indeed the project were real. I also wondered if I was the best candidate for helping fund such a project. I was still a design engineer at Varco International with some music aspirations.

I decided to try. At a minimum I would get to the bottom of the things and find it if this technology was real or not. The business plan Marcus sent to me was well done. From the description of the facilities at LFT and their funding goals I figured that if they were trying to run a con game they were going about it all the wrong way. Pursuing commercial loans, and large companies who would do the most extensive investigations possible before committing money, would be anathema to a con artist.

I called up Bill Glasser. He decided to put us in touch with his friend Robert Schatz who would help us with the project. I called Robert after being introduced by Bill. We set up a conference call with Marcus.

The conference call went well. Everyone was introduced. Marcus talked about the project and his visit. I talked about my technical interest in the project. Robert agreed to take a look at the business plan.

After Robert had time to review the plan we did another conference call. Robert said he had some people in mind that might be interested but we needed to do more due diligence work first.

He would not risk his reputation recommending something fraudulent. Marcus had visited LFT but Marcus is not an engineer and he would be able to ask the tough questions understand what he was looking at. We decided that we would schedule a formal due diligence visit conducted by me. Robert was willing to take my word for it as far as the due diligence went, since he isn't an engineer either. Marcus called up Frank at LFT and we scheduled a visit.

LFT had been through quite a few due diligence visits. They had plenty of supporting documentation and references that we could cross check. Marcus did some of the due diligence. He talked with Dr. Spielvogel at Sandia National Labs to confirm the letter from Sandia.

I planned a short trip. I would visit LFT on Thursday and Friday, and then stay over in Atlanta to visit my brothers for the weekend. I have another younger brother, Mike, who is also in the mortgage business living in Atlanta. We all planned to get together for the weekend after I visited LFT. Marcus was throwing a party at his place.

During my first day at LFT Frank gave us a tour. He showed us all three generations of the LFT reactor including sets of injectors that were melting and giving them trouble. Their first prototype had been stripped down to a cast iron sphere, with injection ports and laser entry points. The top of the sphere was warped from heat damage.

The problem with the injectors was delaying the next test run of the reactor. I was hoping they would be able to run it while I was visiting. They had tried air-cooling the injectors but the way they went about doing so was not sufficient.

A cooling problem is something that can always be solved. It is just a matter of getting a large enough coolant flow. The design requirements of the reactor itself did not inherently prohibit access to the injectors. However, their existing design did limit access, and it was looking to me like the design needed a complete overhaul to make it a commercial product.

I was very tempted to give Frank my resume. They needed some in-house engineering support. The way the design challenges were presented to me, I got the impression Frank thought I would be a good candidate. Frank seemed like a good person to work for.

Instead of offering to help directly I decided to refer Frank to a company called Materials and Electrochemical Research Corporation (MERC). Dr. Raouf Loutfy owns MERC. Dr. Loutfy was one individual that Robert thought could help with funding the project. I had previously studied the MERC web site, and realized that they did rapid prototyping using high temperature ceramics that would be ideal for LFT's reactor.

I was part of this funding team now so I wanted to see that through rather than getting tied up working directly for LFT, no matter how tempting that was.

After the second day of my visit I had a really good picture of what was going on. They were producing a vortex flow of plasma within the reactor and then hitting the tip of that vortex with high energy lasers from multiple directions. This method gave them energy leverage.

Consider the design of a tokamak. All the plasma flows around a donut with a sizeable cross section. It is necessary to heat all the plasma within the reactor chamber to achieve higher temperatures. The LFT reactor only had to act on a pinpoint.

The plasma vortex served to compress everything into that pinpoint. The real key to the design was the electromagnetic containment field used to create the vortex. In nature we see vortex structures at all size scales, from tornadoes up to sunspots and then entire galaxies. Certainly creation of an electromagnetic vortex would be key to a workable fusion design.

LFT was measuring a small amount of helium production in the reactor chamber. This helium count roughly matched the number of atoms that would have to fuse in order to explain the energy gains being measured.

The problem was that the temperature they calculated at the core was less than 1 million degrees Celsius, which is not high enough for fusion according to known physics. Fusion requires two hydrogen ions to merge together. That means the atoms have to overcome enormous forces of electrical repulsion defined by Coulomb's law.

It should require temperatures of over 60 million degrees to ignite a fusion reaction. Even at these temperatures, tokamaks have not been able to produce a sustainable reaction. The moment the heat source is removed a tokamak stops producing fusion reactions.

LFT's reactor had run for hours. In one instance the lasers had been turned off inadvertently and the reaction still continued. LFT's main challenge, besides making the hardware more durable, was developing an understanding of the physics driving the reaction. During each due diligence process they freely admitted that the physics behind the reactor was unknown.

I returned to Austin the following Monday. Marcus and I scheduled another conference call with Robert and briefed him with a little excitement. I was confident that LFT had a viable product, and that this product would eventually lead to a breakthrough in physics.

Robert became more interested in the project. We scheduled another conference call to develop a funding strategy. During that strategy session Marcus went over what he had done so far and what his thoughts and ideas were for funding the project. When he mentioned Mitsubishi, Robert's eyes lit up. Actually, it was a conference call so we couldn't actually see his eyes light up, but we could tell from his response. Robert had spent 3 years living in Japan and working for Nomura Securities. Nomura is like the Japanese equivalent of Fidelity, or some other well-known Investment Company. Robert spoke Japanese fluently.

One of the larger deals Robert worked on involved early stage funding for MERC. That is how he met Dr. Loutfy. Mitsubishi actually funded MERC. Robert had spent well over a year working on closing that deal. The negotiations were high level and included the president of Mitsubishi. In fact, Robert knew for certain that the president of Mitsubishi had a discretionary 100 million-dollar fund intended to explore breakthrough technologies.

Robert said it had been some years since he had been involved in that project. He would need to re-establish those contacts. That would take some time. Robert said he would get back in touch with us later.

Unlike other people Marcus had approached, Robert did not ask for money in advance. Robert worked on projects that personally interested him. The three of us agreed up front that we would all cover our own expenses for travel and things. Should we close the deal each of us would draw upfront expenses from the pot at the end of the rainbow. We would equally share whatever commission was paid for closing the deal.

The payoff for closing a multi million-dollar deal would be very good. The process can be expected to take many months if not years and there is always a chance that the whole thing could fall apart at any moment. That is why Robert was very selective and only worked on things which he enjoyed doing.

Robert contacted Dr. Loutfy to arrange another conference call. Dr. Loutfy was busy and it took a couple weeks to get some time scheduled with him. By then, Dr. Loutfy had read the business plan.

Dr. Loutfy talked over the project with Robert after reading the plan. His main concern was an identification of the physics behind whatever was being accomplished at LFT. I did my best to explain the LFT prototype reactor along with my own ideas regarding leveraging energy to initiate a reaction. As I well knew, that was not a sufficient explanation. Certainly energy was being focused, but that didn't change the fact that they still did not have enough measurable energy to account for fusion.

The other concern that Dr. Loutfy had was that a patent search on LFT only turned up one patent related to waste treatment. Even if LFT had made a brilliant practical breakthrough, it would be very difficult to patent it if they could not explain why it worked. Furthermore, even if they did gain a patent award without a complete explanation of the technology, what was there to prevent another company from making that theoretical breakthrough and putting LFT out of business with a more efficient product?

Regardless of how much market potential fusion represents, investors would expect some kind of assurance that their 50 million dollars would not be wasted.

After all that, Dr. Loutfy agreed to contact Mitsubishi and see if they were in a position to seriously consider the project. That would give us a better idea where we stood. After a couple days Dr. Loutfy reported back to Robert. He said that Mitsubishi would consider sending out a team of scientists to visit LFT if Dr. Loutfy determined that the project merited further investigation.

Marcus, Robert, and myself where excited about that. We were confidant LFT would blow everyone away. We were very close to receiving commissions on a 50 million-dollar deal! All that remained was to set up a visit for Dr. Loutfy with LFT.

That turned out to be much more difficult than my initial visit. By now, LFT was meeting with all branches of the U.S. armed services, people from Sandia Labs, and at least one well-heeled private investor who had heard about the project and flown in from California. The way I saw it, we needed to move quickly and close this deal before someone else did. Neither Marcus, Robert, nor Dr. Loutfy saw things quite the way I did.

Robert was skeptical about anyone writing that check for LFT. He down played the visits at LFT, wondering if those reports where not exaggerated, doubting that anyone was going to fund the company without at least a set of patents in place. Dr Loutfy was new to the project so he had his own personal reservations about the legitimacy of it, just as I had before visiting. Marcus would have liked to close the deal of course. I misread the slow progress Marcus was making getting Dr. Loutfy out to LFT. I thought Marcus was taking a viewpoint similar to Roberts. In reality, Marcus was just having trouble getting time commitments from people.

I wanted to play a more active role in the business side of the project rather than just being a technical liaison. Marcus considered that to be his role in the project, and did not want me calling anyone. In the end, I agreed to leave the business negotiations to Marcus and limit my participation to technical issues.

After more than a month Dr. Loutfy planned a visit to LFT. Dr. Loutfy was doing a trade show in New York and decided to spend an afternoon at LFT on the way back. He would arrange his flight plan for a stop over in Atlanta. Marcus met him at the airport and brought him out to visit LFT during the afternoon.

The visit went very well. Dr. Loutfy decided to do what he could to get Mitsubishi involved in the project. Of course, we would have to split any commission that might be paid between 4 people now to include Dr. Loutfy. We all figured that there would be more than enough to go around if we were able to close this deal.

Over the next couple weeks Dr. Loutfy talked with Mitsubishi. He was ready to schedule the arrival of a team of scientists from Mitsubishi. Not too many days after that Marcus got a call from Tom Beardan. A large U.S. aircraft manufacturer had written a check for 60 million dollars. LFT no longer needed our services.

We were still hopeful that LFT would consider doing an international joint venture. We managed to schedule a conference call with LFT to discuss that. During the conference call we mentioned the name of the large U.S. aircraft manufacturer that Tom had told us about. That was a mistake. We were told that this was very confidential and not to mention that to anyone. So, we stopped mentioning that name, using large U.S. aircraft manufacturer instead.

The military does something called an after action reviews. This process helps soldiers and commanders in the field learn from their victories and mistakes in simulated combat. We had all put significant time and energy into the LFT project, which ended rather abruptly, and it was some consolation to learn why.

LFT's technology was very workable. It mattered who owned it. It even mattered who knew about it. One of the members of the board at LFT had served in World War II. He still had some misgivings regarding Japan. He did not want this technology going over to Japan under any circumstances.

Normally, the process of closing a deal with this large U.S. aircraft manufacturer would have taken much longer than it ended up taking. People were confidant Mitsubishi really was interested. We made it clear that Mitsubishi's funding structure set up by the president of Mitsubishi allowed him to move quickly. If Mitsubishi wanted to, they could give LFT a quick commitment.

There were no consolation prizes or second place compensation. Marcus and myself went back to focusing on our regular jobs wondering what would become of LFT. Barry, the president of LFT, was kind enough to set down with Marcus for an hour thanking him for his help and wishing us all the best.

I continued working at Varco International. I would surf the Internet often, trying to find a clue to the mystery of the LFT reactor. A year later, still working at Varco, I decided to give LFT a call and see how they were doing.

When we parted, Frank and Barry projected having a reactor in operation within 6 months installed in a retrofitted power plant. I hadn't heard anything in the news. The least I was expecting was some sort of press release.

I was reconsidering sending a resume out to LFT.

Working in the oil industry I knew how much harder it was getting to find oil. I knew that mankind was going to have to switch power supplies in not too many years. It is always better to make an orderly transition rather than completely running out of something. At a minimum, the U.S. would be faced with economic collapse in the face of interrupted oil supplies without an alternate energy source in place.

I called up Frank to say Hi. Frank answered the call and said things were going well. LFT had recently built a new facility in New Mexico. I asked about when I could expect to see a press release.

Frank's overriding concern was protection of their technology. I mentioned that the technology really was needed in the public sector, considering our dependence on oil. Frank figured that maybe in 2 or 3 years they could say something about the project. I got the impression that it really wasn't completely in his hands. He was doing the best job that he could, given certain constraints. The department of defense now owned the technology and it wasn't going anywhere else. In fact, even the people working on the project could expect to stay in that position for a very long time.

If I had gone to work for LFT I would have been bound by the same constraints Frank was bound by. With some funding I was confident I could get at least as far as LFT had with the project. I was also making some progress better understanding the process driving the LFT reactor. If I figured out the general case I could produce a better product at a lower cost. I decided to remain independent regarding fusion, at least for the near term.

A few months later I began a thread of conversation over the net with an electrical engineer named Jim Ryan. Jim was working on a nano-fusion idea. Jim's idea was to precisely control the motion of atoms at an atomic size scale. He wanted to etch a little particle accelerator into silicon. I thought that was a novel idea.

We talked over the phone a couple of times. I offered to help Jim put a business plan together to try and raise some money for it. During our conversation he asked me if I had seen the Monolithic Fusion Reactor (MFR) web site. This was a different fusion project that had caught his eye. He wanted to know what I thought about the site.

The MFR is a design concept done by Scott Albright. The site contained an external view of the reactor design, which looked sort of like a large toy ray gun. There were no design details beyond that. What really interested me was the extensive documentation of a new "Apocalyptic" physics, which was the basis of the MFR design.

I wasn't sure what was meant by Apocalyptic. The physics model and the math behind it though, were ingenious. After studying the site for a few days, reading every page in great deal, I realized that this physics model could very well explain what was going in LFT's reactor.

I called up Scott Albright. Scott had been doing his best to get the MFR project off the ground floor, without success. We spent many hours talking about physics and the idea of energy existing in a stable spherical form. In fact, I burned out my cell phone battery more than once listening to Scott.

Scott is a genius, philosopher, and saint all rolled up into one. He really is a good guy. Here he was sitting on the discovery of the century, if not the millennium. He was so far ahead of everyone else that life must have seemed very lonely.

For the past 20 years Scott had been pursuing physics. In just the past few years before I called it had all fallen together for him. He had made some effort to gain Congressional support for a project to demonstrate his ideas. It requires a committee of one's peers to recommend a fusion project for funding in the U.S. There are very few qualified "peers", and as far a Scott could tell they all had their head in the sand. They would look at nothing other than kinetic fusion methods, even though every attempt along this line has failed. By kinetic I mean using heat to produce fusion. Certainly much energy could be released in the form of heat when fusion takes place. That does not mean the fusion core itself has to be hot.

The fusion researchers of the past who have failed to deliver an economical fusion reactor to the American public sit on a committee. There is a bias towards continuing kinetic fusion research. This people tend to insist that kinetic fusion must be right. Doing otherwise would mean that their lives had been wasted pursuing something that could not work. Fusion results in a state of greater order. Fusion starts with two hydrogen atoms doing what they will and ends in one helium atom plus energy. Increasing heat increases the chaos between particles. Kinetic fusion would never be self-sustaining as far as Scott was concerned.

The sun does not produce fusion by heating particles. We know that a strong gravitational field inside the sun brings particles close enough together to fuse. It is a mistake to assume that there is any other way to do fusion. Researchers who do not completely understand gravity have no business wasting money and peoples time trying to do fusion.

Personally, I believe that the committee over seeing U.S. fusion research is nothing more than a façade projecting a public image. I am quite sure that at least one defense contractor already owns fusion technology and does not want to release it to the public. The best way to keep the public guessing would be to set up a congressional committee manned by people who have already failed to create a fusion reactor, and then put congress people in charge of that committee who understand even less than the physics people. All of those people would have almost zero chance of succeeding, and they could honestly and sincerely tell the public that they were doing there best.

After a week or more talking with Scott I knew that he was on the right path and that kinetic fusion would never go beyond where it was today. Huge expensive tokamaks running at high temperatures can only achieve statistical fusion. Small numbers of atoms fuse by accident. Sort of like winning the lottery three times in a row. Two atoms will fuse if they are very close and achieve over lapping quantum states. This will happen now and then by chance in a 60 million degree plasma. One problem is that the energy required to keep the plasma hot has always been greater than the energy released by fusion. Even if someone achieved something slightly over break-even that doesn't mean that there will be any practical way to get useful energy out of a tokamak design.

The path to fusion is not through the chaos of energetic plasma. The creation of stable spheres of energy implies the creation of a gravitational field. That is the path to practical fusion.

Remembering Jim and his nano-fusion idea, I realized that they couldn't both be right. I called Jim and said that I didn't think smashing the atoms together was going to work for fusion. The atoms would just bounce off each other.

I thanked Jim for his help and for referring me to Scott. I decided to try and help Scott with his project.

3

Grand Unification

Scott was looking for funding to build a proof of concept MFR. His reactor design was scalable. His goal was to mass-produce reactors for residential and personal transportation applications. He mentioned there would be a break down of centralized power distribution in the not too distant future, and that people would need these products. Under no circumstances would he accept funding that compromised control of the company.

I considered options for funding Scott's project. Going back to Mitsubishi would have been impossible without having a patented proof of concept to show them. What would we do? Say "Hi! Here is great idea and a design for a fusion reactor, we just need 20 million dollars to build it and patent it." Mitsubishi would just say thanks for the design and build it on their own. Maybe they would send a Christmas card if it all worked.

Raising private capital was the second option. There are barriers that make raising private capital difficult for a start up company with nothing more than an idea.

I believe that the government should not prevent individuals from engaging in a financial transaction of their own free will. That kind of barrier stifles the economy. There are many good ideas out there and hard working people ready to make them a reality given the chance. Pursuing new things is more than half the fun in itself, even if some of those projects fail. Rather than encouraging new business, and giving people the freedom to participate in new ventures, today's system discourages new ventures that would compete with established industries.

First, an individual must be a qualified investor with at least 1 million dollars in liquid assets to legally invest in a private corporation. If an entrepreneur accepts an investment from someone who is not qualified that individual can sue if the project does not make a profit. They might even be able to sue if the venture does make a profit. Certainly there needs to be protection for investors. That protection can be clearly outlined in the original agreement with the investor.

Granted, there will always be risk, that does not mean the government has to stick its nose in and take away the right of millions of people to invest in private ventures.

Second, small business investment corporations are either illegal in most states or virtually impossible to form. The concept of a small business investment corporation is to pool the resources of people who would like to invest in early stage ventures, but are not qualified investors.

Third, it is illegal to publicly advertise private ventures. All ventures have to begin as private ventures. Underwriting an IPO is generally not done without operating history. The company has to have at least 5 million dollars of tangible assets to start trading publicly. The company can go down to negative 5 billion in assets after the IPO. I am just talking about starting private and public ventures. If all public ventures begin as private ventures, and we virtually outlaw private ventures, where is that going to lead us?

It is obvious where this leads us. Over 95 percent of the public are not qualified investors and can only participate in the public stock market. That means that a very small group completely controls what new projects will be started, and also controls the price at which these companies will be sold to the public. Why are things like that? Do you really think the laws regarding private investments were passed purely to prevent fools from being separated from a little money? The public stock market has done that to people in the past as well as any private venture plan could have. At least with a private venture the investor gets to work with the CEO, see the money being spent, and know exactly why things worked or didn't work.

What happened to free enterprise and the spirit of the entrepreneur this country was built on? I believe we are coasting along on the laurels of the past, and need to improve the situation.

Even without dozens of laws discouraging new ventures it would still be very difficult to raise private equity. It is not easy to find private investors.

A few years earlier, Scott raised some money from individuals in Australia. It wasn't very much, but he managed to build a little device in his garage that could generate some fleeting ball lightning. Ball lightning is a natural phenomenon involving a semi-stable sphere of energy. Ball lightning has been reported doing some very interesting things.

A sphere of ball lightning was observed by an entire group of people inside a commercial airliner when the ball lightning floated through the skin of the airframe, moved down the aisle, and exited through the top of the aircraft. No damage was reported. Ball lightning is just a big particle. A particle is nothing more than a spherical solution to Maxwell's equations.

Maxwell's laws resulted in equations, which govern electromagnetic interactions. I am limiting the amount of math I present in this book. I would prefer people to focus on visualization of new concepts first. Once a concept is well visualized the math is nothing more than a tool to define the precise behavior of the model that one has visualized. There will be a need to present some math when I present an expansion of relativity.

Maxwell's laws involve electrical flow and the magnetic fields produced by those flows. People generally apply these equations to a two dimensional linear flow, or two dimensional circular flow. It is possible to apply these equations to a three dimensional spherical flow. This has been neglected in the past because people did not believe that independent spheres of energy existed.

The main discovery for Scott was a recognition that energy can take on two forms. Energy can flow from point to point, or it can exist in a stable spherical form. You might consider this quite obvious; of course a sphere of energy can exist! As obvious as it may sound, this statement is completely new to most everyone in physics these days.

Commonly accepted knowledge, which is true knowledge, has a boundary or cutting edge in a society. In the days of cavemen, that cutting edge might have been forging metal. This concept I am about to present is beyond the cutting edge here in 2003.

Consider an orange. The orange has a core, orange sections, and a skin. The top of the orange has a stem, and the bottom does not. Put the orange in front of you. If you don't have one, you could go get an orange, which might be easier than imagining this. I want to be sure we are looking at the orange from the same orientation.

Slice the orange in half from the stem down, cutting through the center. You should see two cross sections of orange sections. The outer edge of each orange section is the path of energy flow for a sphere of energy in the dipole current mode. When I say dipole current mode I mean a direction of energy flow that results in a magnetic dipole.

This is the mode that most of Earth's magnetosphere is in right now. Magnetic North is one side of the dipole and magnetic South is on the other.

You may have been told that the magnetosphere is caused by an iron core or by magma convection. This is false. There are about a dozen different ways to demonstrate why this is false, some of which will be covered in the next chapter.

For the moment just take my word on that. A breakthrough is something new. That means it flies in the face of the status quo to some degree.

The status quo says that any energy flow or magnetic field has to be caused by something else. Something else like an iron core that loses its magnetic properties at core temperatures. I am talking about a sphere of energy that just exists. This isn't very new. Using an electron microscope we can see that an atom is just a stable sphere of energy. What is new is the idea that stable spheres of energy can exist that are bigger than atoms, and that Earth's magnetosphere is one example. These larger spheres can exist at any size, and at a higher or lower frequency than people expect. These spheres have other interesting properties. For now we will just consider the geometry of energy flow within the sphere and how it changes.

Consider an orange with current flows going along the outer edge of all the orange segments. Looking at spheres of energy in the universe, like the sun's heliosphere or the earth's magnetosphere, scientists have noticed that these magnetospheres oscillate. Oscillation means that the flow of energy changes. Every eleven years the polarity of the entire heliosphere of the sun completely reverses direction. This happens every eleven years like clockwork and no one knows why. Atoms also oscillate like clockwork. In fact, some people use those oscillations to build atomic clocks.

Do people ask why atoms oscillate? Most researchers have not. I guess they just accept that atoms were created that way. That is a good observation. Why have people lost that common sense looking at magnetospheres in this universe that are obviously oscillating spheres of energy? Using Maxwell's laws it is not very hard to calculate the energy that would be released by such a large sphere of energy oscillating by its very nature. Those calculations explain dark matter, solar cycles, magnetic storms, geologic processes, and the fact that we have an expanding universe.

When the sun is midway between changing polarity an event called solar maximum occurs. Solar maximum is characterized by enormous solar flares and energy releases on the sun. Why should these energy releases occur during solar maximum?

Maxwell's fourth law states that a changing current flow induces another current flow in space. It may sound strange that just by one current flow changing we get a whole new current flow just showing up in space. That law is true. It has been validated by experiments. The universe works the way it works, not because we assume it should work a certain way. The universe couldn't care less about people's aversion to creation of energy, or action at a distance.

This is an important point. Knowing that atoms oscillate for billions of years should have caused people to question the unproven law that energy cannot be created. The fact that the universe is expanding should have caused people to question this assumption. Knowing that Einstein himself considered his work incomplete, and no one had presented a grand unification model, should have left the door open for questioning mere assumptions that have not been proven. It is silly to even propose a negative like "energy can not be created". What kind of progress could possibly result from that? The same is true about the statement "energy can not be destroyed".

After all is said and done it still looks to me like energy cannot be destroyed, even though creation of energy is the most common thing going on in this universe. Even so, it would be more accurate to say, "as far as I know energy can not be destroyed, it can change forms". That is accurate. That statement does not discourage other people from trying to figure out how to destroy energy. Maybe someone will figure out how to do that someday and get the Nobel Prize for it.

If the sun's heliosphere is an independent sphere of energy, which oscillates, we can expect these oscillations to release large amounts of energy at the core of the sun. Using Maxwell's laws, it is possible to model a sphere of oscillating energy.

The earth's magnetosphere is ten times the size of the planet. It contains enormous amounts of energy. Even in a stable dipole current mode, changes in the magnetosphere create energy flows at the core. This results in the production of self-similar structures at the core. A self-similar structure means that a big sphere of energy creates smaller energy spheres at the core. We call these self-similar energy spheres neutrons. Maybe that is why people have used the term mother earth.

Does this violate the conservation of energy? The answer is no. Energy is conserved once it is created. The conservation of energy does not say anything about creating energy.

Let's get back to the oscillating orange example. Visualize current flows concentrating energy at the core and creating a magnetic dipole. Scientist know that the Earth's dipole reverses, but up until now no one could say exactly how. Looking at the orange cross section again, the energy can flow clockwise or counter clock wise around the outside edge of each section. The direction it flows determines whether the plus side is at the stem or at the bottom of the orange.

How does the current flow go about changing directions? It would be a discontinuity if the current flowing around the segment edges instantly reversed directions. That is not the way it works. The energy sphere is a 3 dimensional object. The current makes a smooth transition through a new circular current mode in 3D space. Actually, that transition is a little rough resulting in solar maximum and what will soon be called earth maximum. The transition is still not a discontinuity. After spending some time in this new circular current mode, the sphere of energy transitions once again to a reversed dipole mode. Current going the other direction around the orange segments represents the reversed dipole.

I am going to call this transitional current mode the circular current mode. I call it the circular current mode because the magnetic dipole disappears while the sphere of energy is in this transition mode. People wondered why the sun sprouted two North poles during the mid point of the last solar cycle. There is nothing mysterious about that. We are going to need a new orange to demonstrate the circular current mode.

Consider a new orange hanging from a tree. Leave the stem on the tree, and cut off the bottom half of the orange leaving the top half on the tree. Looking at the cross section now, you see a small cross section of all the segments within the skin. The skin is a circular shell holding things together. During the transition phase of an oscillation, the current flows in circles that are concentric to the circle formed by the orange skin. This example with the orange is a simplification of the circular current mode. I reality this circular flow of energy results in a vortex structure like a hurricane or tornado. This structure was clearly imaged by solar probes launched to study the sun during the last solar cycle.

An animation of an oscillating energetic sphere has been produced showing the vector flow of current for a sphere of energy oscillating between current modes. This animation is posted on the book web site:

www.answersforeverything.com

There is a link called "Oscillating Particle" containing the animated gif.

This universe is alive and well. Spheres of energy oscillate continuously producing energy. Planets, suns, and moons grow. The initial size of a magnetosphere determines how fast a celestial body will grow. Jupiter's moon IO is a very good example. IO has many active volcanoes. All heavenly bodies had a magnetosphere in the beginning. There is a cycle of growth, stabilization, and decay.

Continuous oscillations of a magnetosphere over millions of years produce quite a few neutrons at the center of the magnetosphere. A neutron in space decays into a proton and an electron. There we have the three basic particles. A gravity field is produced at the core of every magnetosphere. The gravity of earth is a combination of the monolithic gravity field produced by the magnetosphere, plus the combined gravity of all the atoms that make up the planet.

People have presented some really crazy hollow earth theories. People have said that the planet is hollow, and there are nice cozy villages in the hollow of the planet. Some false explanations presented by people without a technical education can be as strange as the false things presented by modern educated people. The planet is hollow, but it is certainly not cozy. Fusion processes at the core form elements from hydrogen all the way up to uranium. Things cool down to hot magma in the mantle. From there, eruptions to the surface produce islands and new landscapes. Believe it or not, the exact center of the earth's core is as close to absolute zero as things get. The intense gravity pulls particles together, and slows down vibrations. Vibration, and particle motion, is what we call heat.

The model I just presented took me at least 6 or 7 weeks to work out with Scott. Going over the math he presented on his web site helped. He didn't publish all of it. I was shocked when everything started falling together. I finally understood what he meant by "Apocalyptic physics".

I have always had a certain faith in the truths presented in the Bible. Seeing it all integrated with the language of mathematics, logic, and visualization, really changed the way I look at things. Scott considers that his efforts where inspired by God. I believe that. I also believe that God works in everyone's life. In fact, I consider that to be inescapable reality. It takes all kinds of people to build a civilization. I am happy to serve in any role that comes my way. That keeps life interesting.

Fortunately, the earth's magnetosphere does not reverse every 11 years, as does the Sun. The last complete reversal of the earth's magnetosphere occurred almost 800,000 years ago. In the last 200 million years the magnetosphere has reversed 171 times. The time period between reversals on earth is very erratic.

There are a number of possible explanations for the lack of a regular time interval. One thing that is constant though, is the pattern of energy flows during a reversal. If we graph the field strength over time for the magnetosphere, we see a pattern. The dipole strength goes along steady and strong, then the dipole strength drops rapidly to zero. After this transition, it returns rapidly to full strength with reversed polarity.

Over the past 800,000 years the earth's magnetic dipole has been strong and steady. Until 2000 years ago, when it began to drop rapidly. Today it is at about 10 percent of original strength, dropping at an exponential rate. The transition is sectional at first. This results in more hurricanes and tornadoes, which are vortexes caused by sections of the magnetosphere entering a circular current mode.

We are headed for an event that could be called earth maximum. There is more to understand about the physics driving this change. If this earth maximum event were hundreds of years away then there would be no real urgency for the general public as far as understanding things goes. That is not the case. Change is accelerating on this planet at an exponential rate. We have long past dealing with geologic time frames. I would be surprised if we have more than 10 years before all hell breaks loose.

Experts have one job. The general public has an entirely different job. In a democracy big projects ultimately get done because the public supports those projects. Experts do not wield enough votes to make a difference on a national scale. Even though many experts agree with the research I am presenting the ultimate decision to take action or do nothing is in the hands of the public.

That is why I would like everyday people to understand exactly what is going on. That is why the manuscript for this book went through "beta testing" involving many people without college educations to see if the information was presented clearly enough for them to understand. In the past, experts have failed to discover this new information and the media has not been very helpful getting the information in this book to the public. I hope that will change regarding the media. I do not know that it will change, and it may be that only hope is for a grass roots spreading of this information.

There is a lot of false information being promoted in journals, on the web, and by the media regarding science and physics. That false information threatens to lead all of mankind into oblivion unprepared for the future, as it will be.

First I will present a grand unification of natural forces and then a simple and improved version of Einstein's principle of relativity. I will present this as best I can. I hope many people will understand and educate others.

The problem of grand unification is the problem of relating 5 forces observed in nature. These forces are electric, magnetic, gravitational, the strong nuclear force, and the weak nuclear force. The principle of relativity is a very broad statement that seeks to define the nature of the entire universe. Relativity deals with space and time, as opposed to grand unification, which deals with forces.

Gravity is the key to understanding grand unification. Before I get into grand unification and relativity, I would like to tell a story about 2 men and a stone. It helps to clear one's mind before tackling a serious subject like grand unification.

Once upon a time a man was walking in a woods. He was walking rather briskly. As he rounded a corner, he stubbed his toe on a stone. The man cursed and raved. He screamed aloud.

"This stone is not real! This stone is a 3D holographical force projection created by the evil alien race living inside the planet!"

"I know these demons. They go by the name of the druids who worship the iron core! Yes, I have been tricked! But they can't fool me...No! I'll show them a thing or two."

And so the man kicks the large stone with all his strength, really stubbing his other toe.

Another man walks down that same path one day. He stubs his toe on the same stone as he turns the corner. This mans looks down and is impressed by the smooth lines and sparkling light reflecting off the stone in the sun light. The man wonders at the beauty of the stone and conceives a plan.

He decides that these stones would make wonderful pets. They are smooth and petable. One would never have to worry about forgetting to feed or water them before going on vacation. "Yes!" The man says. "I will start my own pet stone business!" "I'll make millions of dollars!" Of course I'll have to find some smaller stones. The cost of shipping would be quite high on those big stones.

The man looks about and finds thousands of similar small stones. A plan forms in his mind. He decides that he will need at least 100 million dollars in first round funding to really get things started right. It takes him a couple weeks to raise the money for the project. All and all things are going well, with one exception. The investors are demanding a say in the day-to-day business operations! They would like to call the product pet rock instead of pet stone. Reluctantly, the new entrepreneur agrees. Everyone is happy. They all make millions selling pet rocks.

The magnetosphere is no different than anything else we stumble upon in nature. Just because it is big, and not quite as solid as a stone, is no reason to assume that it is not a primary object. Today mankind has learned just enough to be dangerous in the field of physics. It is important that everyday people recognize what is at stake and what are options are at the national level.

High yield underground nuclear testing in the 60's generated massive electromagnetic shocks big enough to influence the magnetosphere. It is very likely that these events contributed to an acceleration of the ongoing reversal of Earth's magnetosphere. Large impacts like the meteor that hit the dinosaurs can also accelerate or trigger magnetic reversals.

Nuclear detonations of any yield, at any location inside or outside the planet, only serve to accelerate the process. A planetary oscillation is a natural thing. Asking how to stop it, or how to slow it down, is like asking how to extinguish the sun. Movies like "The Core" are good entertainment but "fixing" the planet's core is only fantasy. There is nothing wrong with it to fix.

Detonating a large bomb near the core could really mess things up possibly leaving the planet uninhabitable for good. I doubt that is even remotely possible given 100,000 years of technological advance. The "unobtanium" talked about in the movie is exactly that, unobtainable. At least we know that during the last 171 reversals life continued. Granted, one mass extinction event saw the destruction of 95 percent of all species.

Scott Albright developed the spherical energy model that I have presented. James Clerk Maxwell developed the math that drives electric and magnetic forces during the oscillation of a sphere of energy. Basically, Scott recognized that a sphere of energy could exist, and solved Maxwell's equations for energy flowing in a spherical pattern. I would like to go over exactly how gravity fits into the picture. That is the answer to grand unification that people have been searching for. It took me years to figure this out after talking with Scott. In fact, it was not until I started writing this book that this fell into place. Maybe that was because it was the one missing piece and I needed it to complete the manuscript. After dealing with grand unification I will present a structural model of the universe. This model improves upon the purely theoretical model presented by Einstein.

Once again, the problem of grand unification is the problem of relating 5 forces observed in nature. These forces are electric, magnetic, gravitational, the strong nuclear force, and the weak nuclear force.

There have been a variety of models presented describing the atom. At first people thought that electrons were little balls flying around in orbit about the nucleus. Then people started doing some testing looking at atoms with electron microscopes. People realized that atoms were fuzzy.

A better model of the atom was put forward involving electron clouds or electron shells. This is correct. However, there is still one misconception regarding the atom. Talking about an electron shell, people assumed that the mass of the electrons resides within the shell. This is not the case.

Really, there is no mass per se. What we perceive, and measure, is a gravitational force. That is all that mass is, the presence of a gravitational field. Recognizing that spheres of energy produce gravitational fields solves grand unification.

When electrons join an electron cloud or shell, they merge into a larger sphere of energy that oscillates as I have described. Spheres of energy are the building blocks of this universe, large or small, with an infinite variety of oscillation frequencies. The gravitational field is strongest at the center of the energy sphere, and then weakens with distance. As the distance to the gravitational source center approaches zero the gravitational force greatly increases. It would be very difficult to measure the gravitational gradient within an atom. It is practical to measure this internal gravitational field looking at man made spheres of energy about the size of a basketball.

Can you guess what really causes the strong and the weak nuclear forces? If you guessed gravity, you are correct. The electron cloud produces a gravitational field with a source at the center of the atom. That field attracts protons and neutrons in the nucleus. The protons are separated in the nucleus by neutrons when possible due to Coulomb forces repelling the like-charged protons.

Larger atoms with more electron shells can hold more particles in the nucleus. Radioactive decay occurs because a nucleus can become unstable. A balance can exist between electrical repulsion and gravitational attraction in the nucleus. Since all the particles in an atom are in constant motion, sometimes a particle manages to escape the balanced nucleus. This results in radioactive decay.

People invented a strong nuclear force to account for the fact that two like charged protons stick together in the nucleus. Next, they invented a weak nuclear force to account for the fact that the strong nuclear force didn't always stick everything together.

That gets rid of two out of five forces, leaving us with the challenge of relating the remaining three. This will require some additional hardware. Possibly a nail, piece of cardboard, and some kind of sphere that we can nail things into. I suppose another orange would work, though it might get a little messy.

For this experiment it does not matter how the orange is oriented. Draw a cross on the piece of cardboard and prepare to nail it into the orange. The nail should go through the point intersecting the two lines of the cross. The X and Y-axis of the cross represent force vectors for electric and magnetic forces. These vectors are tangent to the actual vectors on the surface of the orange. The actual forces curve around the sphere of the orange. Electric and magnetic forces are always at 90 degrees to each other.

It does not matter whether you call the X-axis the line of electric or magnetic force. As the particle oscillates between current modes the electric and magnetic force lines swap between the X and Y-axis. The nail driven into the orange is the cross product of electric and magnetic forces. The nail represents the line of action for gravity, always pulling things towards the center of the sphere.

That is grand unification in a nutshell. This is a three dimensional universe with forces acting along each dimension. It is important to realize that time does not exist as a reality in this universe. Memory is just a sequence of recorded events. The events and the recordings are the only things that are real. It is a mistake to say that something is real when in fact it is not. This universe is just continuously happening. Fortunately, the motion of many things falls within certain legal boundaries defined by quantum mechanics, so there is some order and pre-dictability as the universe goes on continuously happening. It is never completely predictable on either a small scale or a large scale since the universe was created with an element of free will. Truly understanding this life becomes much simpler, and many lies are exposed. For example, since time is not real it could not possi-ble be the cause of anything including aging. The process of aging cannot be inevitable; it must be the result of negative events impacting the body that are not repaired by the body. The body is capable of healing, so there must be a solution to the problem of aging. Time is nothing more than a convenient measure of the rate of motion within a particular frame of reference, whether that frame of refer-ence is a heavenly body, space ship, or artificial gravity field.

"The Universe In A Nut Shell" was a great title for Stephen Hawking's last book. If he had only realized how close the title was to the truth he may have fig-ured it all out.

Stephen Hawking made a great contribution discovering that black holes actually radiate energy. A black hole is what you see when the core of a strong gravitational field is laid bare.

In this galaxy, the only obvious black hole is at the center of the galaxy. Black holes are just a very special case of a gravitational field strong enough to hold even light. Mass does not get sucked into black holes. Energy may merge into the larger sphere of energy that is actually the source of the black hole. The particle does not go into the black hole. The energy of the particle is dispersed and joins the halo of energy surrounding the black hole.

There is a strong electromagnetic field pervading our galaxy, similar to the heliosphere and the earth's magnetosphere. Weaker gravitational fields exist in every particle that has mass. Spheres of energy much larger than atoms can easily produce core gravitational fields strong enough to fuse hydrogen. The force of gravity drops off very quickly moving away from the center of an energetic sphere, especially beyond the boundary of the sphere itself. At the exact center of the sphere the gravitational force is very strong.

There is no danger of black holes eating up the universe. Black holes are the engines of creation in this universe, and responsible for defining the structure of the universe.

Solving relativity is a little more complex than grand unification. It is not overwhelmingly complex. I am going to explain it a couple different ways from Einstein's point of view. Einstein's viewpoint is very theoretical. Then, I will explain it from a simpler point of view bringing absolute space back into the equation. This simpler point of view is an expansion of relativity representing a higher point on the pyramid of physics knowledge.

Einstein claims to be describing the entire universe in the principle of relativity. Actually, he says nothing about the structure of the universe. Possibly he didn't know that the universe had a particular structure.

Einstein's work is still very important and completely valid. Einstein's work gives us the mathematical tools to relate time and space in this universe. None of these things change. I will present is the why behind these things and tie it all into physical structure.

Here is the principle of relativity followed by the constancy of light.

1.   The principle of relativity. There are an infinite number of systems of reference (inertial systems) moving uniformly with respect to each other, in which all physical laws assume the simplest form (originally derived for absolute space or the stationary ether).

2.   The principle of the constancy of light. In all inertial systems the velocity of light has the same value when measured with rods and clocks of the same kind.

The first statement is saying that the universe is made up of an infinite number of objects, some of which are moving faster or slower than others. Einstein then says that the laws of motion, which Newton defined, apply within the scope of each object. Einstein abandons the concepts of absolute space and the stationary ether.

Breaking away from absolute space caused Einstein many years of trouble and criticism. This trouble was not necessary. It is possible to arrive at the same conclusions that Einstein arrived at without abandoning the concept of absolute space. Doing so, one can use a model that explains the structure of the universe as we observe it. Einstein's principle of relativity is weak. Obviously a great deal of structure does exist looking at the universe. All Einstein is saying is that there is an infinity of stuff flying about "every which way but loose".

The constancy of light is an experimental observation that never would have been guessed if someone had not discovered it through experimentation. The constancy of light is a startling experimental observation. We find that no matter how fast different objects are going a particular beam of light always appears to be going at the same speed when viewed from these different objects. Obviously there is something very special about light.

It is going to take some explaining to describe the implications of the constancy of light. The result of that observation is so incredible that it took many years of repeat experiments by dozens of labs before everyone was forced to accept that it was true.

One confusing thing about Einstein's principle of relativity is that he makes an assumption that even he does not know he is making. It really is impossible to know that two objects are actually moving at different speeds without assuming that there is an object at rest with respect to the two moving objects. Oddly enough that viewpoint at rest considering that two objects have relative motion is Einstein himself. Before Einstein could write his equations he had to assume that two object were moving with respect to each other. According to Einstein's principle of relativity there is no way to tell from the objects themselves.

What I am talking about is bringing consciousness into the picture. Logically, consciousness must be at rest with respect to everything else when we start using consciousness to talk about the entire universe, which is what Einstein is trying to do. I would not mention this if it didn't go beyond philosophy.

There is an object at rest with respect to any object moving within this universe, and that object is a very large magnetosphere pervading this universe with a very large black hole at the center of the universe. That is the only explanation for the structure that we observe in the universe.

Pretending we do not know that, let us take a look at what Einstein discovered by using this experimental observation called the constancy of light. The speed of light is 186,000 miles per second.

Here is an example. Say we get ready to turn on a laser that will shine towards a planet 10 light years away. Ten light years away means that it will take a beam of light ten years to arrive at the planet, at which time the people on the planet will see the light appear in the sky. We know that the forward edge of the laser beam is going to take 10 years to reach that planet from earth.

Our friend Tom jumps in a space ship that is capable of achieving a speed of .99 times the speed of light the moment we turn on the laser. .99 times the speed of light is fast. It is not as fast as light though so Tom should always be behind the end of the laser beam. We will find that Tom is indeed always behind the laser beam, but some other interesting things happen during the trip.

Jack is on earth ready to hit the switch that will fire the laser and simultaneously launch Tom instantly to a speed of .99 times the speed of light. Tom is a strong character. He will grin and bear the forces of acceleration.

Jack hits the switch launching Tom towards the planet and following the laser beam. Tom is expecting his journey to last 10.1 years. Tom has a complete library of books to keep himself busy. Tom also has some high tech sensors in his ship. He can measure the forward edge of the laser beam, as well as the distance to celestial bodies like earth and the planet he is approaching.

After Tom reads through an entire set of encyclopedias he decides to check his instruments.

Tom does not know about the principle of relativity or the constancy of light. Tom is shocked when his sensors report that the leading edge of the laser beam is moving away from him at 186,000 miles per second! Tom knows his ship is going .99 times the speed of light. After all, he barely survived the acceleration. He didn't expect to be at a stand still with respect to the forward edge of the laser beam! Tom is worried that his ship stopped and stranded him out in space.

Surely if Tom were chasing another car and that other car was going 100 miles an hour and he where going 99 miles per hour he would only lose ground at the rate of 1 mile per hour. Tom does some quick thinking.

Speed, or velocity, is equal to the distance traveled per unit of time.

$$\text{Velocity} = \frac{\text{Distance}}{\text{Time}}$$

That means that if he has 20 miles to travel to get to work he gets there sooner if he drives faster. Tom is quite sure that velocity, distance, and time are related. The problem is that this light beam measurement has really shaken his confidence about things. Tom doesn't think he is stopped because he can actually see that he is moving ahead looking at known constellations in the sky.

Tom decides to take another reading. He measures the distance between the earth behind him and the destination planet 10 light years away. Now Tom is really surprised. The destination planet is now much less than 1 light year away. That means he is going to arrive 9 years earlier than planned! He will never finish reading the entire library. In fact, it looks like he will arrive in about 6 months.

Tom arrives at the destination planet and the welcoming party greets him. They congratulate him, patting him on the back, saying he did a great job arriving exactly on schedule.

Tom says "wait a minute, I'm nine years early!" Nobody believes Tom. They think he must have set his watch back and gone a little space crazy being the first human to travel in the new ship. For the sake of the story say the people on earth and on the destination planet had a means of instant communication. That is how they know earth launched the ship 10 years ago.

This is the result of the constancy of light. On the two planets, not moving with respect to each other, the light beam takes ten years to cross the distance moving at the speed of light. Inside Tom's ship that same light beam is seen to move at that same speed of light away from the ship according to the constancy of light, which has been tested. Something has to give in the velocity equals distance divided by time equation. Velocity can't give because experiments prove the constancy of light. Therefore, it must be time and distance that change, and this is indeed validated by experiment.

After 6 other people make the same journey they all report arriving early. The instruments recording to the ship's hard drive all confirm the astronaut's reports regarding smaller distances between earth and the destination. The light beam is always moving away at the speed of light. Somebody on the destination planet proposes the constancy of light.

After the constancy of light is accepted someone else comes along and does some algebra. This algebra recognizes that the earth frame of reference, and Tom's ship frame of reference, are both looking at the same beam of light. An equation of motion is written from Tom's frame of reference in the ship that involves the light beam. Another equation of motion is written from earths frame of reference involving the same light beam. The two equations can be manipulated and set equal to each other because they have that same light beam in common.

There are a couple useful ways to solve these equations of motion. One solution compares the perception of time on the ship versus time on earth. We can plug in higher ship velocities and see much different ship time and earth time become. The equations can be solved for the perception of distance aboard the ship versus the perception of distance on earth. These two solution look very much alike, allowing one to plug in different ship velocities and see how time or distance changes.

Tom says, "of course distance and time change as you approach the speed of light! I was there, I know!" Tom thanks the person who did the algebra anyway. Now Tom can calculate exactly how much earlier he is going to arrive based on his speed in relation to the speed of light.

Here is Einstein's formula:

$$\text{Ship\_time} = \text{Earth\_time} \times \sqrt{1 - \frac{\text{Ship\_velocity}^2}{\text{Speed\_of\_light}^2}}$$

The ship time approaches zero as the ship velocity approaches the speed of light. It is interesting to note that the result is an imaginary number if the ship velocity is greater then the speed of light. Math is just a tool. Writing this math formula does not mean it is impossible to go faster than light. Granted, using the chemical engines we have now we will not go faster than light.

As velocity approaches the speed of light the number subtracted from 1 becomes significant. At a ship velocity of .99 times the speed of light we get the square root of .01, which equals a time dilation of .1. That means that ship time equals $1/10^{th}$ that of earth time when traveling at .99 times the speed of light. On Earth 10 years have passed. On the ship only 1 year has passed according to a watch inside the ship. From the viewpoint of earth everything is moving in slow motion inside the ship, if one could see inside the ship from earth.

Why does the universe work like that? The nature of time is one thing that has puzzled people for a very long time. Some people have said that time is 4th dimension. That is not what time is. The reason that time has been such a mystery is that there really isn't any time per se in this universe.

Here is an explanation for what is going on with relativity inside Tom's ship. The motion inside the ship is actually a composite, which also includes the motion of the ship going at .99 times the speed of light relative to this universe. From the viewpoint of earth the ship is still moving slightly slower than the speed of light. Consider net motion as a constant throughout the universe. As one unit of universal "time", or block of motion ticks by, a certain amount of motion is allowed to occur. The motion is what is important. Time is just a concept like mass is just a concept used in place of a gravitational field. Since Tom's ship is going so fast relative to the stationary center of this universe, less motion is available for each tick of Tom's watch in that universal unit of motion. The only reason that time appears to be real is that time is completely defined by memory or recordings of the physical universe. If memory did not exist there could be no perception of time. What is real is that memory of motion taking place relative to a stationary universal center. Time is not real.

Consider all objects in the universe moving with respect to a stationary universal center. Some objects are moving faster than other objects. If a universal unit of motion exists, then any ticking watch will be a composite consisting of the watch tick plus the motion of whatever the person is sitting on holding the watch. Each tick of the watch is motion added to a motion base. That motion base is different on Tom's ship than it is on earth. During each tick of Tom's watch the fast ship he is sitting in is taking up a big chunk of that universal motion slice. In ten years of universal time represented by a constant amount of motion, Tom's watch only gets a chance to do 1 years worth of ticking because he sitting in such a fast ship. The people back on earth have watches ticking on top of the relative motion base of earth itself, so those watches on earth tick more during one block of universal motion.

Hopefully that makes a little sense. If not, it isn't particularly important anyway, unless you plan to start designing faster than light spacecraft. Some things are more important than other things in this book. One very important point is recognizing the exact nature of change that will happen on this planet driven by core energy releases. Another important point is to understand the grand unification model I presented well enough to educate people that don't know about it. That probably includes everyone who has not read this book.

Let us return to the story of Tom. We do not want to leave him stranded on the strange planet.

Tom decides to make the return trip to earth at .99 times the speed of light and visit his friend Jack. Tom arrives on earth and notices Jack has some gray hair now. Jack is 20 years older. Of course, Tom knows this is due to the constancy of light.

Tom walks up to Jack, slaps him on the back, and jokes about how Jack will never fit in at the spring break party he is planning to go to this evening. Tom says Jack is old and has gained much weight. Jack hits Tom with a judo chop to the side of the head, shouting, "Who do you think you are! Get out of my office!"

Jack and everyone else on earth have not learned about the constancy of light. Tom was the only one brave enough to make the return trip and go through that rough initial acceleration again. Tom is sad about having lost his friend Jack.

How can Tom prove he is who he says he is? How can anyone for that matter really prove what has actually occurred in these two separate frames of reference?

It sounds like I am playing the devil's advocate questioning the very truth of Tom's existence. After all, I started the story sending poor Tom off at .99 times the speed of light with instant acceleration.

Einstein did not take anything for granted developing the principal of relativity. I am just being a little obstinate about it.

The only way to really prove what has happened between these two frames of reference is to admit a real third frame of reference at rest with respect to Jack, Tom, and every other frame of reference in the universe.

That of course is a sphere of energy as large as the entire universe defining space and monitoring motion. Spheres of energy can exist at any size and have no problem merging or over lapping each other.

What this assertion comes down to is an argument over infinite space versus finite space. Let us put aside philosophy and consider the technical merits of both assertions, which do contradict each other.

Einstein is saying there are an infinite number of inertial reference frames. Einstein places no boundaries on space. I am saying that a finite number of particles exist in this universe, and that space has an edge.

Physics is based on proven physical law. A statement asserting that there is an infinity of something is impossible to prove. As such, that statement should have no place in physics.

An individual could try and prove there are an infinite number of particles. They could go out and count for ten years, then return with a very big number. I could always say keep counting till you find them all. I could only win the argument. Logically, there is no way I could lose.

The same is true for space. Someone could look out with a big telescope and say space is infinite because they don't see an edge. I can always say look farther until you do.

A tenable position regarding the universe would be a statement that the universe consists of N particles with N-1 relative relationships existing between them. These relationships would be defined by the principle of relativity. When N-1 particles move relative to each other, there would have to be an Nth particle, which is at rest with respect to all other particles.

Even if there wasn't one big particle that began this universe, holding space together, math could still be written to support absolute space. With N particles in motion it would be possible to select one particle and arbitrarily consider it at rest. Then one could write N-1 relative relationships, while at the same time having a means of making absolute measurements using the particle determined to be at rest.

The truth is there is one large particle that defines the boundaries of space for this universe. Looking out as far as we can see with the Hubble space telescope, one large structure stands out. In addition to billions of galaxies and galactic clusters, there is a structure that has been called the great wall. The great wall is a sheet in space with a denser concentration of galaxies than the surrounding space. This sheet looks like a section of a spiral arm, not unlike the spiral arms that form the Milky Way galaxy.

If the Hubble telescope could look out much farther, it would see that the great wall really is part of a spiral arm. This universe might look like a big version of the Milky Way Galaxy, with individual galaxies in spiral arms rather than stars. When the Hubble telescope looks at 360 degrees of sky, things are roughly uniform, so it more likely the universe is in a dipole current mode that is a more uniform distribution of energy. The great wall that Hubble sees could be more like a regular cloud in the universe, rather than a hurricane like spiral arm structure.

This begs the question; "How far up does it go from there?" Is this organized universe containing galaxies, just one part of an even larger structure? Researchers estimate that the Milky Way contains about 100 billion stars, and that the universe contains about 100 billion galaxies. This observation is evidence of similar structures at larger and larger scales. Let us call the organized structure above galaxies the universe. If there is an even bigger organized structure with each particle in it being a universe of 100 billion galaxies we can call that a mega universe.

I would bet a Coke that if someone looked up past "our universe" as it seen by NASA with 100 billion galaxies, a mega universe would not be found. That is my opinion. I figure I am safe for some time on that Coke bet. One would have to look very far indeed to find something an order of magnitude larger than 100 billion galaxies. One hundred billion galaxies is certainly enough to keep mankind busy for a long time

That is the principle of relativity in a nutshell.

Einstein was not satisfied with the principle of relativity. Einstein went down a path of research without the advantage of experimental data like the constancy of light to back him up. His goal was to bring gravity into the picture. The result of his work along this line is called General Relativity, which is still debated. Experimental data is lacking regarding general relativity, and Einstein was not able to arrive a unification of forces when he talked about gravity. Even so, general relativity deserves some attention. I will not try to untangle the math of general relativity. Just because someone writes some math does not mean that math is an accurate model of the universe.

All experiments confirm the principle of relativity. Going forward from here I will rely on imaginary experiments to visualize some of the concepts of general relativity and gravity. For the most part, experiments that will confirm or deny what I say have not yet been done. One exception is observations related to the propagation of gravitational forces. Gravity is known to act instantaneously across any distance within this universe.

The principle of relativity implies that the only way to slow down Tom's watch is to get him moving close to the speed of light relative to his friend Jack on earth. This is not the only way to create this effect. If Tom were to step inside a sphere of energy on earth with a gravitational field at its center, Tom's watch would slow down. A gravitational field can also be considered a time field.

Traveling forward in time is a realistic proposition. Tom would not be torn apart stepping into a properly created gravitational field. The movie "Time Machine" is a good visualization of such an experiment. That movie also made a good point by implying that it is not possible to go backwards in time. That is one negative assertion I will bet on. It is only possible to speed up or slow down motion in one's vicinity. Greatly slowing down motion in one's vicinity makes it look like one is traveling forward in time.

The reason why it is not possible to go backwards in time is because there is nothing real about time itself. Independent versions of this entire universe do not exist for every instant in the past, affecting things in the present. Memories of the past exist, but these are just energy recordings that exist in the present and are subject to alteration. This universe exists in the present and is based on motion in space.

We know that a moving electrical charge creates a magnetic field. Tom's ship could be considered a single particle because it is all moving at the same speed.

Tom's ship is actually producing a halo as he accelerates towards the speed of light. This sphere of energy increases the strength of the gravitational field in the vicinity of Tom's ship, which is the actual mechanism behind Tom's watch slowing down. This effect also increases the apparent mass of Tom's ship. The individual atoms that make up the ship are not changing. Mass is nothing more than a gravitational field. The mass of Tom's fast moving ship now includes the gravitational field of all the atoms plus the gravitational field of the halo pervading the ship. Einstein managed to arrive at this conclusion via an enormous amount of math, probably without actually visualizing what was happening. How Einstein managed to figure this out is anyone's guess.

The increased mass of Tom's ship equals the small gravitational field produced by each ship atom combined with a gravitational field surrounding the ship produced by acceleration. The math of general relativity can be used to calculate the strength of the gravitational field produced by a rapidly moving object.

That is general relativity in a nutshell.

A visualization of what is really happening in the universe is just as important as mathematics. Math is just a tool to approximate real events. The math of general relativity allows people to calculate exactly how much the mass of Tom's ship will increase. This kind of information is valuable to an engineer building a space ship.

I am writing to give people a general understanding of this universe. I want most people to understand that there is good reason to believe the near term predictions that I present for this planet.

If most people in the United States do not understand these models I expect most everyone would walk into oblivion not knowing what hit them until it was far to late. There are signs in the environment that give people a vague premonition that major change is on the horizon. That is not enough to motivate people towards action. A detail understanding of just how much energy is going to be released, and exactly what the environment is going to be like for some years, is necessary to motivate people.

Every new idea or break through encounters resistance from the existing status quo. It takes a great deal of courage, and more than a little knowledge, for people to debate against people with PhD's who might cling to models of the past like iron core models for examples. Certainly some experts agree with the model presented in this book and I have posted some reviews at www. answersforeverything.com. Only time can tell how quickly this work will be accepted. Truth always wins out eventually, but it can take time and that is what concerns me. I cannot say exactly how much time this planet has until the environment precludes advanced preparation. Everything that I see tells me that it is not that many years, certainly much less time than I would like.

They say that if something is repeated three times it makes a bigger impression. The future of this planet is in the hands of the people who read this book. Books sell or do not sell based on word of mouth more than anything else. I have done what I can writing this book. I will also promote it full time until 2008 or until enough people have been reached to achieve the political goals I have mentioned. Even so, one person can only knock on so many doors.

Let's take a look at three practical applications of energetic spheres. These will be interstellar communication, a practical approach to fusion, and travel faster than the speed of light. I present these applications just to give people the idea that these things are possible. I am not trying to present design level detailed drawings, just enough to show that these things are not impossible.

Interstellar communication involves sending information much faster than the speed of light. Stars are many light years apart. It would require a method of communication that is instantaneous, or so close to instantaneous that light looks like a turtle in comparison.

The fact that gravitational forces act instantaneously across the entire universe makes interstellar communication theoretically possible. The only reason people have not taken advantage of this has been a lack of technology to create gravitational fields. Actually, that isn't entirely accurate. I am sure the B1 bomber uses electricity to create gravitational lift, it is more accurate to say that these things have not been publicly available.

If gravitational field changes are experienced without delay across any distance, the solution is to have a transmitter that varies the gravitational field and then a receiver to detect those changes. Of course, that is a huge technical challenge with all the other gravitational sources around that would mask the signal. It is still theoretically possible.

This behavior of gravitational fields is quite amazing, but true. It is proven by the fact the people can accurately predict the motion of heavenly bodies only by assuming gravity acts instantly across any distance. Here is an example.

If two stars where separated by ten light years they would each experience a force of attraction caused by the gravity of each one acting on the other. If one star were quickly pulled away the other star would instantly experience a reduced attractive force. Some people shudder at the thought of this.

How can one star know the other star is there ten light years away? People shudder because they have a problem with action at a distance. If people don't like action at a distance they can go build their own universe without action at a distance. Living in this universe, people will have to live with action at a distance. The structure of this entire universe is based on non-local cause and effect. That means that gravitational changes in a large black hole at the center of this universe would instantly effect things on this planet and everywhere else in the universe. On the flip side, changes in gravity field on earth instantly affect everything else in the universe, however small that effect may. The universe is made of N particles experiences N-1 instantaneous relationships.

That is the theory of interstellar communication. In practice, it will be found that manipulation of gravitational fields can produce effects that propagate through the universe much faster than light but slower than the instantaneous effect of gravity.

A practical approach to fusion is next. People have assumed that the only thing that has to be done to achieve fusion is overcoming electromagnetic repulsion between like charged hydrogen ions. This is the smash the atoms together theory of fusion.

Overcoming electromagnetic repulsion is just one piece of the puzzle. Even that piece of the puzzle has been misunderstood. Coulomb's law says that like charged particles repel, which is correct, until particles get so close that quantum interactions begin to influence the two particles. A friend of mine named Merchello Spiega helped me improve my presentation of this.

Merchello Spiega is an accountant working for Visiosonic. Visiosonic produces PCDJ, the leading professional DJ software. Merchello played a lot of football in school. I walked into Merchello's office saying I wanted to present a generalization of Coulomb's law for him and see if it would make sense to an accountant.

Merchello took offense to my insinuation that accountants who majored in football might not understand things any better than the average person. He agreed to hear me out anyway.

I explained that liked charged particles normally repel each other. When they get really close though, about 1 particle diameter away, they begin to attract each other.

Merchello raised up his two fists and began moving them together slowly with a grin on his face. As his fists came close together Merchello said; "Yeah, I can see that. This force that acts at a distance can't work the same when they get really close together because it gets sort of entangled."

I hadn't actually mentioned anything yet about quantum mechanics. I really didn't need to. Merchello had the concept and he expressed it as well as anyone might. I told Merchello if he could get credit for that he could win a Nobel Prize. Merchello said he always wanted to be a scientist.

Here are the quantum mechanical details of this phenomenon. Like charged particles experience attractive interaction when they share a common overlapping De Broglie wavelength. Prince Louis de Broglie is famous for his work in the field of wave mechanics. A particle like a proton is never a stationary object. There is a wave function defined by quantum mechanics associated with every particle. The two particles must be at rest with respect to each other for this attraction to occur.

What this means is that people have been over estimating the amount of force that is necessary to bring two protons together for fusion. It still takes a lot of force, but not as much as was previously thought. That is important, considering people have used this force calculation to raise tens of billions of dollars to build tokamaks reaching temperatures approaching 100 million degrees.

There is one more logical error behind the failure of a generation of scientists to deliver an economical fusion reactor to the American people. Say you knew that pool balls on a pool table would experience a force of attraction and merge if you got them close enough to each other.

How would you go about merging the two pool balls? Would you pick up your number 20 cue and fire one pool ball into the other as fast as you could? You might. That attractive force would not get much chance to operate as the greater force of impact sends one ball flying off the table.

A better plan would be to grab hold of those pool balls, regardless of how much they where repelling each other, and hold them both together so that they could merge. Fusion driven by a gravitational field is the only practical way to do fusion. Not only do the particles have to be close together to fuse; both particles have to be at rest with respect to each other. A large gravitational field reduces the rate of motion within that field, getting both particles closer to being at rest.

Kinetic fusion is a waste of time and money. Why have people reported some success with tokamaks? The reason is that they have heated trillions upon trillions of particles to such high temperatures that they achieved statistical fusion. Statistical fusion will go down in history as an icon representing how far mankind will go down a wrong path insisting that it can be done.

The ions in the plasma have enough energy to get close enough, but they are almost never at rest with respect to each other when they get close. Two particles near each other win the lottery three times in a row. We get a little fusion. If the energy release ever breaks even with the energy required to heat all those particles, I would be surprised. Considering that one of these tokamaks costs billions to build, and the reaction will never be self-sustaining, it is not worth pursuing any further.

The correct solution is fusion by gravitational fields produced by spheres of energy, or a vortex of energy.

Fusion and communication improvements are well within the reach of existing technologies. Building a real space ship is a much bigger project. I would not pursue such a project until more urgent projects are completed. It would be difficult to even estimate the effort without trying different approaches over a period of years or decades at great expense. Here is a theoretical approach.

Consider what happens when a ship is accelerated towards light speed. It is true that the mass of the ship increases, which is the principal argument against faster than light travel.

How is the mass actually increasing? Since we know that there is no such thing as mass per se, just gravitational fields created by electromagnetic flows, we can answer this question.

The increase in the mass of the ship can only have two sources. Either energy is added to the electron clouds of every single atom in the accelerated object, or there is a background gravitational field created by acceleration.

If the electron clouds of each atom changed that would wreak havoc with chemistry. Fortunately, observations of fast moving particles like cosmic rays do not indicate that this is happening. Tom's ship accelerating towards light speed is picking up a sort of halo.

If it were possible to create a counter gravitational field something could be done about this halo, which is increasing the apparent mass of the ship.

One example of a counter gravitational field on earth involves hurricanes and tornadoes. The drop in air pressure at the center is caused by a gravitational effect on particles. Electric force flowing in circles does a good job of whipping up some wind. If we were to look, we would notice a sort of hole in the magnetosphere in the vicinity of these events.

I found a good story from some fighter pilots who flew a squadron through and into the eye of a hurricane. Those pilots discovered that a magnetic compass does not work properly in a hurricane.

It is true that convection, and being over warm water, contributes to storms. Water, especially warm water, has a unique ability to store energy. Those things are only part of the picture. A drone would have to be built to fly all the way through the center of a hurricane to test this. The drone would need electric, magnetic, and gravitational sensors. In the absence of tests like that, all I can do is suggest that somebody take a look. Governments do that kind of research because there is no immediate commercial benefit. There is the benefit of public knowledge, which is one reason we pay taxes.

I hope this discussion gives Star Trek fans hope for the future.

The destruction of the second space shuttle underscores the need for better space technology. In the near term I would go with fusion drives. Not long after that we should be able to field drives working directly with gravity.

Chemical drives are just not meant for space. Low energy chemical drives force extreme design compromises. The shuttle faces extreme heat and friction on reentry because we cannot carry enough fuel to get into orbit and then back down at a more controlled pace.

The astronauts of NASA are very brave people. I am sure they will continue to go into space as necessary with chemical rockets no matter what the risk. It would be a sad thing indeed if other departments had better technology and failed to share it with NASA.

Fusion power is not a dream on the horizon. Fusion power is available. The first workable fusion design just didn't go public. Nor will any other until the public insists upon it. The LFT design I described was primitive and expensive. There are far more economical fusion designs that will produce greater energy gains.

Chemical rockets are fine for satellites. I would hold off on the manned missions until the technology issue is straightened out at the national level.

There is one more formula that I would like to cover which has become popular. This formula is written in two different forms:

Energy = Mass × Velocity^2

Energy = Mass × Velocity_of_light^2

The first version of the formula uses the velocity of a moving object. This formula represents the amount of energy imparted to an object by acceleration. A car going faster does more damage hitting a brick wall.

The second formula represents the total energy of any object at rest. Mass is nothing more than a gravity effect produced by energy. We measure gravity and call that mass. The fact that energy equals mass times the velocity of light squared gives you an idea of how much energy is an atom. It take a bunch of densely packed energy to create an atom. That is all it is though, dense energy. Now, consider how much energy might exist in a big atom like the magnetosphere. The results of releasing some of that energy will make that formula very real to people in the future.

Consider one pool ball at rest and another pool ball that you are about to hit with a cue. When both pool balls are at rest they both have the same energy. When you hit a pool ball you transfer energy from your cue to the pool ball. The moving pool ball now has energy equal to the mass of the pool ball times the velocity you hit it with squared.

Now consider the energy of a pool ball at rest. We can figure out how much energy was used to create the pool ball by multiplying the mass of the object times the speed of light squared. That is a lot of energy. The magnetosphere surrounding Earth is ten times larger than the planet. This sphere of energy is capable of trapping energy radiated by the sun over long periods of time. There is no shortage of energy at the planet's core to use for creating neutron particles in a quantity sufficient to make enough magma to grow the diameter of the entire planet. This growth occurs in spurts.

There is a concept called entropy that is poorly understood today. An individual that blindly accepts entropy as truth could use this concept unwittingly as an argument against the creation of energy. There really isn't any arguing with the creation of energy in this universe. Even so, the concept of entropy has become so prevalent, reaching even into popular fiction works, that it deserves mentioning. Entropy began as a simple observation of a chemical thermodynamic system. People designing thermodynamic systems discovered that after putting 1000 joules of heat energy into a thermodynamic system, it was then impossible to get all of the 1000 joules of heat energy back out of the system. Knowing that energy could not be destroyed, another explanation was needed, and the explanation presented was called entropy. Entropy is the concept that a little of the energy thermodynamic system goes into some kind of chaotic or possibly inert state from which the energy cannot be recovered. This is false. Under the right conditions smaller particles can merge into larger particles. Heat is chaos of motion that is the opposite of order. Out of this chaos of smaller particles a few electrons manage to merge into larger spheres of energy. Essentially, this universe tends towards order not chaos. Particles that are larger than atoms are not recognized by chemistry today, so the formation of these larger particles results in vague explanations like the concept of entropy promoted in chemistry today.

After chemists made this interesting discovery called entropy, philosophers in the fields of fiction and physics then embraced the new concept. The concept of entropy was generalized, and some creative people decided that all matter and energy in the universe must be degrading towards inert uniformity. People writing a dictionary thought that was a great new definition of entropy and so they wrote it down in the dictionary. After being written in dictionaries, the majority of people started believing that the universe was degrading towards inert uniformity. This new definition of entropy is a false claim that is trying to describe the structure of the entire universe. Einstein plus all the best physics people who came after him have tried to define the structure of the universe without success. At least those people on the cutting edge of physics acknowledge their failure to embrace universal structure, leaving the door open for future discovery. This entropy claim regarding the universe is just an unsubstantiated chemistry spin off, not true physics.

Order and expansion are the dominant forces in this universe. From the chaos of heat a little order is born. Chemists are blind to the form of that order, just as physics has been blind to the source of 90 percent of all gravitational forces in this universe. The source of that order is simply small independent magnetospheres, or larger ones like the one surrounding the very planet that we all walk on today. Every electrical circuit that includes an earth ground is actually grounded to the earth's magnetosphere. Not understanding the magnetosphere, people who design power grids wonder how a "weird" natural event could take 27 power plants off line across the Northeast U.S. If lightning was the cause of that power outage, at a time when the weather was not unusually severe, why doesn't the whole power grid go down every time a big storm rolls in? The reason is that the power grid is very well protected from lightning strikes. The power grid is not well protected from major changes within the magnetosphere itself, like magnetic storms for example. A typical magnetic storm is not the only thing that the magnetosphere can dish out to disrupt a power grid grounded to the magnetosphere. The recent outage in the Northeast did not occur during a large magnetic storm, which is usually caused by solar flares. Relatively large independent spheres of energy can form around the planets core and be released to the surface. This has not been very common in the past. Sometimes this phenomenon is called Saint Elmo's fire. In the future, such energy releases will become stronger and increasingly common until the magnetosphere has completed a reversal cycle. The solution is distributed local power generators that do not require an earth ground.

So, there is no valid concept of entropy that can be applied to the entire universe, and the phenomenon called entropy observed in thermodynamics has bee been poorly interpreted. This universe will continue to expand into the void forever. That is how the universe was designed, and that is how it will be.

When the public demands an answer related to something that there is no good answer for, or related to something that authorities do not want to provide an answer for, authorities have only one option. That option is fabrication of an explanation that the majority will accept. For the Iraq war, the answer was weapons of mass destruction. The reality of the situation was much more complex, and could not be explained to anyone other than a full time intelligence analyst accurately. The U.S. is now dependent on foreign oil, regardless of how that came to be. Prevention of near term economic collapse required control of a significant strategic oil reserve. Buying that oil from Saddam Hussein would have put a lot of money in the hands of a dictator prone to violence. Personally, I believe the human rights issue in Iraq would have been the best justification, followed by possible weapons of mass destruction. The administration should have admitted that the U.S. needed the oil, and was willing to pay the people of Iraq for the oil, but was not willing to pay a dictator prone to violence with a record of using weapons of weapons destruction. In all fairness, it is remotely possible that the weapons of mass destruction did exist but were hidden or smuggled out by Saddam, and remain in the wrong hands to this day. The whole mess could have been avoided if the U.S. had moved away from oil a long time ago. Things will only get worse in this area, compromising ethics to satisfy basic needs, if the U.S. does not take action now to provide for the energy needs of the nation without oil.

The public wanted a quick explanation for the power grid going down in the North East. What was the public told by the media? The public understands lightning, and that lightning can cause disruptions, and so for lack of a better explanation lightning was chosen as the best explanation to feed the public. At first, some radio stations mentioned a "weird" natural phenomenon causing the event. I could hear the trepidation in the voice of the radio announcer reporting that. Announcing a "weird" natural phenomenon causes public confusion and panic, especially after movies like "The Core". The radio stations were quickly told to stop that and tell the people it was just a little lightning. It is not a violation of the freedom of speech for the owners of a media company to decided what will be reported. Actually, it would be a violation of the freedom of speech trying to force media companies to report a particular position, whether that position be true or not. Public panic doesn't help anyone, so until the entire public can be educated the lightning explanation is just fine.

What is the current state and rate of change of Earth's magnetosphere? That is a very important question. I will cover this in the next chapter in detail. I spent months of effort trying to disprove the timing and the conclusions I present for this planet.

It has been 800,000 years since the last reversal of the magnetosphere. The strength of the magnetosphere has been dropping exponential for the past 2000 years. It is now at less than 10 percent in some locations. When the magnetosphere shifts to a new current mode it will do so very quickly, in a matter of weeks as evidenced by magnetic traces in old lava flows. It is important to realize that this change is exponential not linear, and that the weaker the existing current mode gets the greater the chance of a single solar flare pushing it immediately to a transition.

4

Empirical Data

Empirical data includes those things that can be measured by anyone. Given similar environmental conditions measurements produced by different observers should be very similar. Any model that does not predict honestly gathered empirical data should be questioned, if not simply discarded. The empirical data is reality. A model is just a tool to help a person understand the universe, or some part of it.

This search for empirical data began as an attempt to shed as much doubt as possible on the idea that we had anything serious to worry about in the near future regarding the environment. I do not consider a little global warming, or gradual flooding of some coastlines, anything really serious. That kind of gradual change is easy to adapt to. Catastrophic change, like the whole magnetosphere changing modes in less than ten years and exposing the entire planet to very deadly solar flares, is the kind of thing I wanted to shed doubt on.

All the research I did only served to elevate my concern regarding near term catastrophic environmental change. The Internet provides an unbiased wealth of information for anyone to review and crosscheck information. I encourage people to do this. You will find that most of things covered in this book are considered either partial or complete mysteries within the scientific community. One day a new scientific community will be built on the information presented in this book. On that day and going forward people will find it hard to believe the level of ignorance that existed prior to this publication. Hind sight is always 20/20. As simple as these things are it is sort of like saying a few lottery numbers are a simple combination. The right numbers are simple, after the drawing.

These dozen or so relevant topics and observations were gathered over many years. The model defining the motion of bubbles for example is quite simple, and obvious knowing what one is looking at observing bubbles in motion. Once again, that same motion can look very confusing not understanding the basic rules of interaction. Bubbles have been studied extensively and have remained a mystery up until now.

Let's start with solar cycles.

Solar Cycles

The sun's entire heliosphere goes through a reversal cycle every 11 years. The heliosphere is just a different name for a larger magnetosphere surrounding a star. The heliosphere is as large as the solar system. The heliosphere reverses rapidly, and very frequently, compared to the earth's magnetosphere.

No one has put forward an explanation for the mechanism behind reversals of the heliosphere. Apparently there is a limit to what improbable explanations people will invent. The sun is known to be mostly hydrogen so saying that the sun has a huge iron core with enormous magma currents would contradict known fact. So, the phenomenon stands as one of the great mysteries of physics.

Solar maximum lasts for about three years. Solar maximum is a period during which the sun is between dipole modes. Solar flares and other energy releases increase 10 fold during this time.

In the middle of the reversal cycle the sun's dipole disappears and lines of force in the heliosphere change from a dipole configuration into a pinwheel shape that looks sort of like a huge hurricane in the heliosphere centered on the sun. This observation has also been a complete mystery. The sun is simply moving through a transitional circular current mode.

This solar maximum transition last for 3–4 years before it moves out of the transition and into a reversed dipole mode. Circular current modes can last for any amount of time, depending on the exact characteristics of the os cillating particle. The Milky Way galaxy is a spiral galaxy right now because the magnetosphere of the galaxy has been in a circular current mode for some time. It is difficult to precisely predict, but I have reason to believe that the earth's magnetosphere will also take about 3–4 to transition out of a circular current mode. This time frame is important because the physical aspects of life on this planet, with the exception of seeds and some ocean species, will not survive in such an environment. That is why this planet has seen so many mass extinctions. Fortunately there is more to life than the physical nature we observe. The spiritual nature of life is very real and capable of "rapidly" advancing evolution after a mass extinction. Rapidly in terms of geologic time frames.

During the last solar cycle probes sent to study the sun noticed that in the middle of solar maximum the sun sprouted two north poles. This surprised people, as did the pinwheel shaped energy configuration in the equitorial plane of the sun's heliosphere. There is a misconception in physics today that magnetic monopoles cannot exist. Once again, people should be careful about asserting negatives. Monopoles can exist. What is an electron if not a negative mono pole? The North Pole is negatively charged, and a South Pole is positively charged. When the sun sprouted two North Poles the entire heliosphere could have been compared to a big electron vortex. A little searching on the internet for data on the last solar cycle will turn up this information. I would rather have people search authoritive sites of their own choosing to validate the data I present. That is the only way to provide objective third party data. If any information exists to contradict what I say this would also be found doing an objective search. In my searching the only contradictory ideas I have found include the whole iron core idea.

People believe that atoms are spheres of energy and that atoms oscillate by their very nature. Until someone proves otherwise people should accept the fact that the sun's heliosphere is also a sphere of energy oscillating by it's very nature. This is the only plausible explanation, and is in fact truth as will be become all to clear in time.

This whole universe was created using oscillating spheres of energy. Some of these spheres give forth light, like the sun. Some give forth magma via fusion at the core. Magma cools at the planet's surface creating the planet. Jupiter's moon IO is very much like earth was in the early days.

Space Weather Now is a good web site to view the progress of the current solar cycle. Space Weather Now is a public site with data on the current solar cycle plus other educational materials. The next solar cycle will peak around 2012. The year 2012 has stirred up some interest as it relates to Mayan prophecy.

Mayan prophecy does not count as empirical data. What does count is the fact that Mayan calendars are based on the real motion of the solar system through the galaxy. The Mayans understood the motion of the solar system as it wobbles in and out of the galactic plane. It turns out that we are on the verge of crossing the galactic plane. The significance of this is not apparent without understanding the models that I have presented. The entire galaxy pervades space with an electromagnetic field. The galactic plane is a crossing point in that field. The solar system can be expected to experience higher than usual rates of electromagnetic change crossing the galactic plane. It is not a problem of the solar system running into dust clouds as some people have been concerned about. The rate of changing electromagnetic fields is most important as it relates to Maxwell's forth law. The Mayans predicted 2012 as the end of their world. They had good reason to make such a prediction. It isn't the "end" of the world. It is a major change and it will require advance survival system technology for mankind to endure this transition. Not many years beyond this rapid change the planet will be refreshed environmentally and the magnetosphere will return to full strength. A full strength magnetosphere will be a great health benefit.

The solar cycle of the sun clearly demonstrates the power and energy of an oscillating monolithic particle. Solar flares are emitted from the sun during solar maximum that would otherwise engulf the planet if not blocked by the magnetosphere. Looking at what happens to the heliosphere during a reversal cycle accurately represents the same cycle that the earth's magnetosphere will go through during a reversal.

Magnetosphere Reversals

Two thousand years ago Earth's magnetic dipole was at maximum strength. It is now at about 10 percent strength on average decreasing at an exponential rate. Vortexes have formed at the North and South poles contributing to the holes in the ozone layer. Northern lights are moving farther south.

The Northern lights are beautiful. The lights are caused by high-energy particles emitted by the sun and impacting the atmosphere. When solar emissions are not completely stopped by the magnetosphere we get Northern lights. Northern lights will move farther and farther south until the lights change into solar flares exploding in the sky unblocked by a dipole configuration of the magnetosphere.

Geologic research confirms 171 reversals of the earth's magnetosphere during the past 200 million years. Lava cooling along ridges at the bottom of the Atlantic records the orientation of the magnetosphere during a reversal. As the lava cools, metallic grains orient toward magnetic North. Satellite data, and sampling of these ridges, defines the number of previous reversals.

Sometimes people doing research uncover results that are strange and unpredicted. A higher level of knowledge will always explain strange and mysterious observations. Monolithic particle physics sheds light on the mysterious aspects of research done by NASA, geologists, and other scientists. Geologic data gathered from the ocean floor is no exception.

It is possible to determine how quickly the magnetosphere went from a normal dipole configuration with a north pole, to a circular current mode with no specific north pole. Knowing the rate at which lava cools, and seeing changes in metallic grain orientation, we know that in as little as three weeks the north pole can move down to the equator and vanish as a single pole. It would not be wise to wait for this rapid motion of the pole to begin before preparing for a very extreme environment. That would leave only three weeks to prepare all of mankind to endure forces capable of destroying all of mankind. Even if people doubt the models I present, I would ask "do you feel lucky?"

That event involving the rapid disappearance of magnetic north will signal the beginning of any panic that is going to occur. Before that day change will be little more extreme than it is today. That is why a magnetic reversal is dangerous to a civilization. There is little or no warning prior to catastrophic change, and a civilization has to reach a level of knowledge understanding the entire universe in detail before the event can be predicted.

A very important question for planning purposes would be "how long will this transition take?" In other words, how long before the protective mode of the magnetosphere returns, and things calm down in the earthquake and volcano department. That is a very difficult question that I don't think anyone can answer precisely. We know solar maximum lasts for a few years. My best guess would be 3 or 4 years. I would not count on that though. I would be best to build systems that are designed for long-term survival in an extreme environment. I would also build industry suited for an extreme environment so people are able to go about their daily business and move society forward in the face of a challenging environment.

Hurricanes and Tornadoes

Hurricanes and tornadoes are driven by sections of Earth's magnetosphere transitioning into a circular current mode. The structure of a hurricane is not unlike the structure of the Milky Way galaxy.

This can be validated by precisely measuring electromagnetic and gravitational fields within a hurricane or tornado. The eye of a hurricane is an area of low pressure. Flying a very sensitive gravity sensor through a hurricane, into the eye, and out the other side, would be a good experiment.

Sensor equipment exists to get very accurate measurements of gravitational fields, electric fields, and magnetic fields. I have searched everywhere for this kind of data for hurricanes without success. Apparently people are afraid to fly directly through hurricanes.

That fear is well founded. I did find a report from an air force test squadron that flew through a hurricane. A nearby air base was threatened by the hurricane. The pilots were needed to get planes off the ground at the other base. The commander who ordered the mission was reprimanded, but it did result in some interesting reports.

Every pilot and co-pilot reported navigational equipment failure. Their radio compasses gave erratic results, sometimes wandering 180 degrees off.

It would take some public money to do this test. Test drones should be built with a precise sensor package covering gravity, electric, and magnetic fields. These drones should be flown into the hurricane, through the eye, and out the other side at various altitudes. These precision results should convince even the most skeptical people.

I would expect the results to indicate that a gravitational field anomaly contributes to the eye of the hurricane, and that winds are driven to higher speeds by an electric field. A magnetic field would also be detected which is super imposed on the earth's dipole field. In fact, all the fields I mention would be super imposed existing fields. For this reason the sensors used would have to be calibrated and capable of high resolution measurements. It doesn't take much of a field change over an area the size of a continent to have a large local effect.

The vast majority of weather is driven by the magnetosphere. People try to predict weather with some success by looking at the inertia of weather in motion combined with rules of thumb. It is true that temperature variations contribute to weather. Neglecting the contribution of the magnetosphere makes it difficult to accurately forecast weather. People still do a fair job. It is possible to write software called neural net programs. These programs use empirical data as input. The software uses data from the past to predict how the weather will be in the future. This works, as long as the driving force behind the weather does not change significantly.

Energetic weather phenomena like storms, lightning strikes, hurricanes, and tornados will increase as the magnetosphere approaches a transition. A hurricane represents a large part of the magnetosphere transitioning into a circular current. As the planet moves towards the whole magnetosphere transitioning into a circular current mode hurricanes could easily decrease for a little while as events like tornados increase. Such transitions are complex, just as solar maximum is complex, so predicting exact events is very difficult. The one constant will be an increase in the net energy that the magnetosphere releases into the planet over time.

Expanding Earth Models

The Atlantic Ocean floor from the United States all the way to Europe is a series of under water mountain ridges. Geologic testing indicates that each mountain ridge represents a polarity reversal of the magnetosphere.

When lava cools grains of iron are polarized in the direction of the magnetic field that existed at that time. The ages of the mountain ridges are older the closer the ridges are to the shorelines. Two hundred million years ago the United States was connected to Europe.

If all the continents were pieced together like a puzzle they would fit together almost perfectly on a sphere roughly 40 percent smaller than the current diameter of earth. Making this observation, people have put forth expanding earth models for almost 100 years.

Expanding earth models have been scorned for one reason. None of the people proposing an expanding earth model could say why, or how, such a change in the earth's diameter could happen. Instead of accepting an expanding earth model that is backed by empirical data, people decided that the planet was lopsided 200 million years ago. The lopsided theory involves another theory called subduction, which is not backed by empirical data. The idea of Pangea was put forward. This theory says that the earth had a continent on one side while the entire other side and more was ocean.

As North America moved away from Europe the Atlantic Ocean was created. Subduction claims that the Pacific Ocean got smaller to accommodate expansion of the Atlantic by plates sliding underneath each other. There is no physical evidence to support subduction in the Pacific. Geologists are not physicists. Geologists should keep an open mind until grand unification is set in stone.

Not understanding the physics that drive planetary change, geologists needed something to write in textbooks. Subduction theory was the best thing going at the time. After all, some explanation had to be written in the textbooks, and no one could justify an expanding earth model.

The earth has expanded. At the beginning of creation the earth was little more than a magnetosphere in orbit around the sun. The gravitational core of the earth's magnetosphere attracted more mass, and oscillations of the magnetosphere released particles that all combined to eventually built the planet.

What will the landscape look like after a magnetic reversal? Geologic change happens in spurts. The sun is much larger than earth and we see structures on the solar surface the size of the Earth changing in days. Normally, the earth changes much slower than the sun, except during a reversal. Within the life span of one generation the landscape of Earth will change dramatically. Saying precisely how it will change in advance is difficult.

Theoretically, using one of the world's fastest super computers a model of the planet could be built. The government of Japan owns a fast earth simulation computer. The magnetosphere could be modeled as a single oscillating particle transitioning into a new current mode. A gravitational field would be modeled at the core, producing a stream of neutrons. Some of the neutrons would decay into a proton and electron. Other neutrons would join with protons and electrons, creating atoms. All atoms are made up of protons, neutrons, and electrons. Higher elements are formed by fusion of lighter elements. The forces of fusion and particle production at the core produce magma. As pressure builds at the core some of that magma is pushed to the surface.

Geologists have estimated that the mantle contains at least as much water as that contained in all the world's oceans. Actually, as much as five times the water in all the world's oceans, according to one estimate done by a researcher in Japan. During a reversal some percentage of this water flows to the surface expanding the oceans.

Combining melting ice caps, existing ocean water displaced by underwater volcanoes, and water arising from the mantle, we could begin to model landscape changes. The effect of surface volcanoes and earthquakes would be added to the simulation. Those detailed calculations have not been done yet and that would take time plus a whole lot of money.

There is a futurist by the name of Gordon Scallion. Gordon has done some interesting work. Gordon claims to have spiritual visions of the future. Gordon created a future map of earth's landscape after a pole shift, which he believes will happen in the not too distant future.

The map includes an overlay of existing country boundaries on top of future geography. For the U.S. the map also includes state boundaries. I am not saying that Gordon's map is very accurate. In fact, there are several things that I disagree with just looking at the future North American topography.

I do not expect large land masses to form off the eastern or western coast of the US. I would expect some volcanic islands, but not small continents. I would not expect the ocean to extend past the continental divide in the U.S. Up close to the continental divide, maybe, but not past it. Gordon's map is a starting point in an area of research that has not begun. Knowing that it is possible to model the future landscape of this planet, and believing that civilizations exist in this galaxy capable of building such a model, I would not be surprised if there is some truth in Gordon's visions of the future. Of course it would be more accurate to just do the simulation. Relying on spiritual inspiration introduces a great deal of ambiguity.

Here is a link to Gordon's site: www.matrixinstitute.com

Being a futurist Gordon tries to predict events down to the year. People judge him on the accuracy of those predictions. I view the future as a set of weighted chances. The lower the planet's field strength gets the more likely that a particular shock will trigger a rapid transition. I will not try to predict the exact timing or duration of a pole shift. I will say that Earth is well on it's way to a rapid transition, so much so that any additional years mankind has to recognize and prepare for this events are gifts from God.

More than one futurist, including ancient cultures like the Mayans, predict that 2012 will herald great planetary changes. For a 3,000-year-old civilization like the Mayans those people knew more than they are given credit for. I suppose if a Mayan priest or the person who wrote down the book of Revelation were here today they might have said, "we told you so". Unfortunately, knowing such an event is going to happen is of very little practical use. The technology that spins off from the models I present is necessary to do anything practical in terms of preparation. Without advanced energy and life technology the only solution is enormous stockpiles of non-renewables. That kind of solution can help no more than some thousands out of billions.

Gordon's map is one scenario representing the scope of rapid change that can be expected. Oceans will certainly rise. Some shorelines will submerge and new ones will be created. Gordon's map is realistic because it models a consistent rise in ocean levels combined with movement along known fault lines.

Magnetic Jerks

During 1969 a great jerk occurred in the earth's magnetosphere. A jerk is a very rapid change in the position of magnetic north. Magnetic North is constantly drifting. Lately, the motion of magnetic North has been accelerating. No explanation has been put forward for magnetic jerks.

People who believe the iron core theory cannot explain magnetic jerks. Obviously it is not possible for the entire mantle and core to rapidly accelerate then slow back down in less than one hour. Knowing the laws of inertia people are forced to place magnetic jerks in the "unsolved mysteries" category.

The very existence of magnetic jerks should have cause people to throw the whole iron core theory out the window. No actual physical evidence for an iron core or rapid magma convection exists. Having no evidence is in itself good reason to throw out a model. Having contrary evidence, one has to question the integrity of people standing behind an iron core theory. Einstein did not believe the iron core theory. One cause of misinformation is pressure to write complete textbooks. Academic textbook publishers market their books as being the leading authority on a particular subject. Admitting one does not know does not sell textbooks. False information gains momentum after being published in a textbook.

It is not possible to advance science without questioning current understanding. Einstein failed math in school. Einstein worked at a patent office. Einstein was not a professor, physics major, or textbook writer. Even so, Einstein did make one of the greatest leaps forward in physics by questioning the beliefs of his day. That trait will always be the hallmark of advance. A PhD does not always convey truth. If you think about it, a PhD means that an individual has spent a decade studying the status quo for the most part, and maybe a little time trying to make some unique contribution that must be related to the status quo if that contribution is going to get a passing grade for graduation. A major advance in understanding that displaces most existing understanding is just not something that going to happen in a classroom. Classrooms are cluttered with old ways of thinking. Besides not teaching things on the cutting edge, classrooms generally lag years behind industrial technology. The most important thing a person learns in school is not so much useful data, but rather the ability to learn how to learn. The degree confers a sort of expert validation so that the individual is expected to accomplish certain things and therefore goes out and gets those things done one way or the other.

The magnetosphere is best modeled as a single oscillating particle with quantum effects. I have only touched on quantum effects. I like to think of quantum effects as free will for particles. I suppose it could even be divine intervention talking about celestial bodies. A celestial body can be viewed as a large complex particle. Quantum mechanics is a science of probabilities. Without getting to deep into quantum mechanics, what it all gets down to is that things like magnetic jerks can happen if the magnetosphere is viewed as a large particle.

Although quantum effects could cause a magnetic jerk, I do not believe that was the source for the global jerk of the magnetosphere in 1969. During the late 1960's the United States and Russia were playing games to see who could detonate the largest hydrogen bomb underground. Each country knew how big a bomb the other country had by measuring the global seismic waves produced by the detonation.

The electromagnetic shock waves produced by a hydrogen bomb are significant. I believe that this testing contributed to the magnetic jerk in 1969. These tests may have accelerated planetary change. I know these detonations influence the magnetosphere. I have not tried to model the exact impact on the magnetosphere for a particular nuclear device.

The nuclear test ban treaty was signed shortly after this global magnetic jerk. I believe scientists and politicians who were involved in this testing recognized that they had done something very bad. Some scientists may have suspected these tests had changed the entire magnetosphere. Fortunately, the nuclear test ban treated was signed. India and Pakistan did not sign the treaty. India and Pakistan tested a dozen nuclear weapons at the beginning of 1998. 1998 marked an acceleration of global climate change. The trend continues. Certainly 400 tornados within 10 days was quite a record during 2003. These days all time weather records are falling shortly after they are set. Unfortunately this is barely the beginning of extreme environmental change.

Recognizing the true extent of the risks carried by nuclear tests on this planet, I would expect politicians to take test bans and non-proliferation more seriously. Unfortunately, the impact of previous tests cannot be undone. The transition of the magnetosphere could be viewed as an avalanche caused by the weight of one to many snowflakes. In my opinion, the current state of the magnetosphere is such that we will not see another solar maximum pass before the magnetosphere moves completely into a circular current mode. There is a risk that further nuclear detonations could begin a rapid transition before any effort can be made to prepare for it, therefore an even greater priority should be placed on preventing such detonation. That is why the war on terrorism and non-proliferation is so important.

Mass Extinctions

Many mass extinction events have occurred on this planet in the past. Volcanic activity is cited as the most probable cause for every one of these mass extinction events. It is true that an asteroid hit the planet about the time the dinosaurs became extinct. It is also true that the large asteroid impact could have triggered a reversal killing just as many dinosaurs as the meteor. A rapid change in climate would have been difficult for the large dinosaurs to adapt to.

I found many interesting books and publications researching mass extinctions. A book called "The Vital Vastness" tries to explain mass extinctions. The author presents something called a field electrodynamic earth model.

I can tell that the author sees a glimmer of truth. He is really stretching it though when he tries to quantify his model. A glimmer of truth is not enough to open people's eyes. Without knowing the whole truth people tend to focus on the holes in whatever new theory is presented.

A complete verifiable model is necessary to gain broad acceptance. Nothing less accurate than the precision with which Newton, Maxwell, and Einstein approached their work will do.

I would prefer not to put to much attention on mass extinctions of the past. It is enough to know that each one of these mass extinctions matches up in time with a reversal of the magnetosphere. Certainly it would be a good idea to put together some kind of national ark program to safe guard things like seeds and different animals to restore those species if it becomes necessary. The U.S. has no such program at this time, which in itself speaks volumes for either arrogance, over confidence, or just a lack of concern regarding the future of mankind on this planet. Even ignoring environmental concerns, mankind has spent 100's of billions on weapons that could destroy everything. Why not spend a small fraction of that on some kind of life ark in case someone manages to use those weapons? The answer is that people have voted to only build weapons of mass destruction, and not build anything to preserve life. It would be wrong to blame politicians. Each individual is responsible for putting politicians in office, money is secondary, people are capable of organizing and contacting others at an exponential rate via word of mouth. As hard as it might be to accept, politicians are doing nothing other than representing the people who put them in office. If the general public failed to organize a word of mouth campaign for a politician, and a special interest group had to step in and help with an expensive media campaign, guess who the politician must serve to get reelected?

I have often wished the information could have been presented earlier. Looking back at my life though, I do not consider any of it wasted time arriving at this point. It has taken time to discover things, and present information in a way that most everyone can understand. I am just thankful that everything did finally come together, and I dedicate future time towards delivering this information to people.

Lava Flows

Lava flows record magnetic field orientation while the lava cools. It is possible to calculate how fast the orientation of the magnetosphere changes since lava cools at a known rate. A 90 degree change in magnetic field orientation has been recorded in cooling lava, indicating that the North Pole can move down the equator essentially disappearing in a matter of weeks.

The fact that most of this planet is planet is molten lava is proof in itself that there is an enormous heat source at the core of this planet. If the planet had no heat source, and if the planet was billions of years old, you can bet that the planet would have cooled solid by now radiating heat into the cold of space for billions of years. A thermodynamic analysis will prove that.

The truth is that this planet and the sun are not billions of years old. This planet is tens of billions of years old, if not hundreds of billions. The sun is hundreds of billions of years old, if not trillions of years old. This universe is much older than people have guessed. This is an expanding universe. The universe is actually growing, adding new matter every second. Recognizing this, a more accurate age for the universe can be calculated. One would have to estimate the size and number of original magnetospheres filling the void at the moment of creation. That number could very well have been one. Rather than a single big bang, the universe has been continuously expanding from the result of billions upon billions of small bangs stretched out over many billions of years.

In this modern age of science and technology the Bible has often been considered less than relevant in the field of physics and matters of creation. History will recognize that the Bible was actually to far ahead of its time to be understood during these early days of technical advance. In all fairness, the Bible was written during a time when language itself lacked the vocabulary to better express ultimate truth.

Jupiter's moon IO

Jupiter has a moon called IO. Active volcanoes cover IO's surface. What could be creating all that heat in the cold of space so far away from the sun? The people of NASA have pondered this question and arrived at a conclusion.

The strong gravitational field of Jupiter must be squeezing IO, creating friction inside of IO, which melts the interior resulting in active volcanoes covering the surface of IO. This explanation is not a joke. Someone actually published this explanation and called it gravitational friction heating. People will believe anything not having any better explanation.

How fast would you have to squeeze and unsqueeze a big rock to create enough friction to melt the rock? I would guess the rock would shatter before it melted. There is no shortage of moons in orbit about planets, and planets in orbit around suns in this solar system. So, why did someone suddenly invent gravitational frictional heating? If this were a real phenomenon, which it isn't even vaguely, then it would exist in other lunar or planetary bodies.

Have people drilled down into IO to see what is really going on? No, they have not. Has anyone done a complete finite element analysis of the moon IO to support this sudden new explanation for volcanoes? No.

These assertions are believed because of the position of the person presenting them. It has little to do with common sense or substantiated scientific method. Sometimes pressure for answers causes people to accept wrong answers. I can imagine the press must have had a field day with the discovery of IO and all the volcanoes way out in the cold of space near Jupiter. Out of 100 scientists who honestly said they did not know maybe one person wanted to be credited with a publication and put forward this gravitational friction idea.

Gravity does have something to do with it. The gravitational core of IO is a fusion power source creating magma inside IO and pushing that magma to the surface.

IO is at the center of a very large hole in Jupiter's magnetosphere. No one has ventured a guess about what causes this hole. Flying over the pole of IO a NASA probe recorded what sounded like an electromagnetic power plant of sorts. This was not detected at a distance from the pole. The magnetosphere of IO is in a circular current mode so NASA does not recognize it as a typical magnetosphere. NASA just sees a big hole in Jupiter's magnetosphere and wonders what's up with that.

IO is not unlike the early days of earth. It is a large sphere of energy spending time in a circular current mode building up IO.

Depending on the exact size of that hole in Jupiter's magnetosphere, IO may become the size of earth many millions of years from now. It is no accident that all heavenly bodies in this solar system, which are not gas giants, show signs of early volcanic activity. That includes the moon. Looking up at the moon higher elevations produced by lava flows are clearly visible. Volcanic activity is the mechanism of planetary formation. The idea that a bunch of rocks orbiting the sun decided to merge together all on their own and create a beautiful planet is just not believable.

The idea that a ball of magma suddenly appeared, and then cooled over time, is just as unlikely. If this were the case we would not see an increase in volcanic activity on earth 5 billion years after the ball of magma started cooling. Scientists know the earth is at least 5 billion years old. Actually it is much older than that. If there is no heat source at the earth's core, and the planet has been cooling all that time, shouldn't it be done cooling? Certainly it should, or there should at least be a thick enough crust to contain any remaining molten magma. Actually, all it takes is a little heat transfer analysis to prove that the core of the planet must be a major heat source. Once again, having no explanation for the source of a planetary or lunar heat source people ignore facts and present odd explanations.

All these things I mention do not go unnoticed by people. I have stumbled across dozens of web sites proposing different models for the earth. One web site proposed a hollow earth model with a small star at the center of the planet. That idea is not far from the truth. The author just did not have physics or products to back up his idea. A correct idea is fine. Proof and a complete mathematical model are usually needed to gain public acceptance of a new idea.

The Galactic Bubble

Our solar system orbits the Milky Way galaxy. The galactic space through which we travel does influence the sun and planet Earth. For some millions of years the solar system has traveled through a pristine vacuum called the Galactic Bubble. As if our solar system were emerging from a womb, 2000 years ago the solar system reached the edge of this bubble. Leaving this Galactic Bubble the space through which the solar system travels contains a little more dust and other particles. We could expect hitting this edge to generate some shock, which may have started a transition on Earth 2000 years ago. The Galactic Bubble itself is a low-density electromagnetic sphere that could influence the solar system as it transitions out of that sphere.

An article describing the Galactic Bubble is linked to at www.answersforeverything.com

Buried in the 20 or more pages of this article, the author makes a comment relating earth changes beginning 2000 years ago to approaching the edge of this Galactic Bubble. Two thousand years ago the magnetosphere of earth began the reversal cycle we are about to experience. The average strength began dropping from roughly 100 percent. Looking at graphs of field strength prior to each reversal of the past we know that these drops always precede reversals. The rate of decline will continue to increase until a new current mode is achieved. It is more complex than the simple 3D monolithic particle animation on the web site. The result will be similar to solar maximum, replacing sunspots with areas of increased turbulent weather. Earthquakes and volcanoes will also increase, releasing enough ash to block the sun.

The sky

Perhaps the most obvious signs heralding change are found in the sky. Today's sky is very different compared to 20 or 30 years ago. Many people in their 30's or 40's can recognize changes in the weather over the past couple decades. Grid like striations appear in the sky today. I am talking about patterns appearing in the clouds that did not exist 40 or 50 years ago. By grid like striations, I mean evenly spaced rows of clouds with spaces between them. Take a close look at the sky. You may see wispy wave-like formations in high altitude clouds at the edge of space. The sky has changed over the past few decades. Northern lights are beginning to appear further south than they have in the past. The frequency of lightning during storms has also increased.

Cloud formations and lightning are driven by the magnetosphere. The Northern Lights are beautiful displays caused by particles raining down on the atmosphere. The particles are not completely stopped by the magnetosphere.

A vortex forming in the upper atmosphere causes the hole in the ozone layer. A vortex flow is typical for a magnetosphere entering a circular current mode. Scientists were surprised to find a vortex over the pole of Mars. Mars has a weak magnetosphere. The magnetosphere of Mars is in a circular current mode so scientists did not take it into account planning Mars missions. This vortex in the thin atmosphere of Mars over the poles caused NASA to adjust the flight path of a probe recently sent to study Mars.

Bubbles

Water bubbles are actually electromagnetic spheres. Fill a bathtub with water and splash the water around. You will see some bubbles on the surface. Watch the bubbles carefully.

Small bubbles are attracted to larger bubbles. The smaller bubbles accelerate as they get very close to the larger bubbles. Sometimes the bubbles merge, usually the smaller bubbles stick to the larger bubbles much like atoms stick together in molecules.

There is a layer of water one molecule thick which traps the bubble. Water has a unique ability trapping high frequency electromagnetic spheres. Usually these spheres pass through solid materials quite easily. The sun emits enough electromagnetic spheres larger than atoms to fill the atmosphere with them, plus the space around the planet. These are extremely low density objects which so seldom interact with anything else they are not noticed.

Even though these spheres are difficult to detect they exist in great abundance in this universe. Space is full of them. This explains the mystery of the missing solar neutrinos and dark matter. These missing neutrinos are just larger and less dense than people expect so they are being missed when people study the sun searching for them. The same is true about dark matter. We know that 90 percent of the universe is unaccounted for. Unaccounted for means that we observe certain gravitational effects, but we do not see nearly enough mass in the universe to account for all the gravitational effects.

Just as smaller bubbles tend to merge into larger bubbles, smaller electromagnetic spheres tend to merge into larger spheres. This could eventually lead to the formation of magnetospheres around stars. Just as we find electron clouds at certain distances around atoms, we find planets in preferred orbits around stars. Once some magnetospheres form in orbit around stars, smaller spheres emitted from the sun tend to join those magnetospheres in preferred orbits.

Bubbles are extremely light, being only one molecule thick. Bubbles move on the surface of water with very little resistance to motion. You will see that as smaller bubbles get very close to larger bubbles the smaller bubble starts moving faster. Gravitational forces between the two bubbles cause this acceleration.

The fact that existing science cannot account for 90 percent of the universe says that if these scientists had to take a test regarding the universe they would fail that test. People tend to be overly impressed by PhD's and Nobel Prizes. In the grand scheme of things the truth really is quite simple. People are impressed because in a world full of confusion by a few people who are either a little less confused, or pretending to be less confused. So, these people win awards relative to the existing level of knowledge, not relative to ultimate truth. That is ok since technology does need to progress on a step-by-step basis. If 100 points were a perfect score on a test concerning this universe, a Noble Prize winner in 2002 would score something like 10 points while the general public scored 3 points. The information in this book should bring the general public score on that universe test up to about 80 percent. The remaining 20 percent involves engineering details. That final 20 percent can get very complex. Even so, the basic operation and structure of this universe is still the most important thing to understand.

Empirical conclusions

All of this information is presented to allow people to make their own decision regarding the future and what should be done about it. This data is just a sampling. More important than data is existing products developed using monolithic particle physics. For example, the crystal products that will be covered in the rejuvenation chapter are pervaded by larger energy fields working to raise the pH of water without the use of a consumable chemical reaction. These products work day after day, decade after decade, and they will never wear out just as an atom never wears out. A higher energy crystal has also been tested that releases enough oxygen from water to produce a form of liquid hydrogen at room temperature that burns as well as gasoline, but without any pollution.

The energy density potential of a sphere of energy is virtually unlimited. At a very high energy density a sphere of energy does tend to expand.

One reason for publishing this book is to educate the public prior to the release of technology that will shift the power structure of this civilization. Right now, the general public is influenced by unseen ropes of economic slavery binding people to a reliance on large centralized systems and high maintenance products. Marketing an energy product that threatens that status quo involves real risk. If a million people understand that these things are possible, and are able to play around with lower energy crystals and experience rejuvenation, that reduces the risk. Eventually, some bright person could bring that higher energy technology to market if the efforts of partners and myself are some how blocked. It would be foolish to think a technology destined to put a trillion dollar industry out of business could be released without some kind of fight. It would take an entire army to defend against a threatened trillion-dollar interest. It is very easy to eliminate a new technology when only a few people understand it, and more importantly while the public remains ignorant of any better way of doing things. In the past people have made the mistake of trying to keep new technology all to themselves. So, if these ideas make sense, and the rejuvenation products work as I say they do, I hope people will help promote these ideas. Individuals looking out only for themselves will always fail. You can bet that people who now control this civilization from an economic point of view work together as a team. Even if you don't agree with how that team decides to run things, it is at least a team rather than a few lone wolfs, and that team is winning at this time. It will take a much better and larger team to change the existing global power structure. A trillion dollars can buy many friends who will sell their soul for a piece of bread. After all, today many people don't even consider that souls exist anyway.

After doing all this research I built a web site focusing on the technical details of grand unification and the implication for planet earth. I paid for press releases to go out to thousands of media outlets. Apparently my press releases were ignored. I did not receive coverage from a single media outlet.

Maybe the media considered the whole thing just another doomsday alarmist press release. At the same time that my press release was not run, a dozen relatively crazy and unsubstantiated alarmist theories were run regarding global warming, weapons of mass destruction, the possibility of broad scale terrorist actions, asteroid impacts, flows of magma suddenly stopping in an iron core earth, etc. Could it be that a large number of falsehoods or half-truths are promoted making it very difficult for people to see what is really happening?

The problem with the media could have been that that the media was not able to communicate anything worth printing, being limited to a one-page press release. After failing with my press release I did not immediately think about writing a book. I just considered that the press was biased and there was no way that I was going to get through to the media.

I decided to try a different approach.

I planned to achieve financial success through my music software. Those finances would allow me to focus on music. With enough money and time I could hire a band. I could lease a large concert hall and break into the music business on my own. Then, I would have an audience. I would also be doing music. That may have influenced my decision to take this round about path.

It turned out that some things I discovered while pursuing music are directly related to the future of civilization. Things like rejuvenation, real economic solutions, and the importance of politics. So, the time was not wasted.

Any serious solution to the challenges of the future must involve economics and politics. Things like economics and politics are not like laws of physics. Economics and politics are based on the combined beliefs and actions of millions of people. These things change day to day unlike physics that is ultimate truth that never changes regardless of anyone's opinion. Engaging in large-scale economic or political projects involves understanding all the people involved, and how those people are changing. I began to learn about these things trying to raise money for my Maestro music project.

5

Fund Raising

Getting back to life, I was still working at Varco International with aspirations of pursuing a music career or business. After some success at the NAMM show in LA I was confident that with some funding the Maestro software that I wrote would do well in the music software market. I decided to raise money from private investors. In order to do this legally I needed to put together a placement memorandum for a regulation D private offering. I also needed a detailed business plan.

I hired a lawyer as a consultant who had experience doing private placements. I received a template that I could use to do most of the work on the placement memorandum myself.

The legal documents, business plan, and market research filled a 200-page binder. This took me three months to put together. I printed 1000 copies of a professional box on a high end press working with a printer in Austin. I decided to send out my business plan, along with a sample of the product, to as many financial groups and people as I could find. Some groups wanted an executive summary first, followed up with a complete package if they were interested in seeing more.

In addition to putting all the paper work together I read whatever books I could find on the subject of raising private capital. I found one very good book, "Angel Financing", by Gerald Benjamin. Gerald also helps people raise money for projects. I sent Gerald an executive summary. A couple weeks later, Gerald himself called, saying he would like to take a look at my project. I was excited about that. I sent him out a package.

I also got a call back from Global Capital Securities Corporation (GCAP). GCAP was a larger firm, and one of the very few that worked with early stage ventures. GCAP was one of the first firms I sent a package out to so they received everything in my first submission.

Clifford Morgan at GCAP called after receiving my plan and product sample. He talked with me for some time, mentioning that his firm had raised 5 million dollars for a music software company called Visiosonic in a private placement. Clifford expected to take Visiosonic public soon. He thought I could work out some kind of deal with Visiosonic, integrating my product with Visiosonic's product line.

The date was August 15th 2001. I had a little vacation time left at Varco, but not very much. I decided to schedule two short fund raising trips. The first trip would be to visit GCAP in New York City where I would stay in Manhattan. The second trip would be to visit Gerald Benjamin at International Capital Resources (ICR) in San Francisco California.

I arrived in New York City on a Wednesday, August 29th 2001. GCAP was located about an hour's drive outside the City in Melville, New York. The hotel I rented was close to the airport on Manhattan. I was flying in late, so I booked a room in New York rather than Melville. I found a reasonably priced room, which was a surprise being right in the middle of New York. They had an Internet special for advance bookings that I took advantage of.

I brought a Korg Z1 keyboard with me, along with a portable Fender sound system and my Dell laptop. I planned to do a demonstration of the software at GCAP. My appointment was at 11:00 am on Thursday. My return flight was not until Sunday so I would have a couple days to see New York.

I set my alarm at the hotel in downtown New York, not many blocks away from the twin towers and the Nasdaq stock exchange. Early the next morning I was on my way to Melville. I always allow plenty of extra time for travel, especially if I am not familiar with a particular area. Leaving at 8:00 am, I figured I would have plenty of time to get to Melville by 11:00. It was only about 40 miles away.

I managed to get off Manhattan without being caught in gridlock. I expected traffic to be busy. The main road going from the heart of the city out to Melville turned out to be much busier than the traffic downtown. I missed the wide-open roads of Texas.

Moving along at about 15 or 20 miles per hour I figured I might just arrive in time. The traffic started getting worse rather than better. I pulled out my map to try and find some secondary roads that could get me to Melville quicker.

I found a route that I thought would be better, and pulled off at the next exit. It was a little better at first, but it starting looking like other people had the same idea. The closer I got to merging back on to the main road, the busier the traffic.

While merging, I was forced into the breakdown lane. I must have picked up a nail or something. I got a flat tire and ended up changing it in the break down lane in my nice new suit.

I called GCAP and let the secretary know that I would be a little late. After changing the tire, I rolled into Melville a little past 11:00 and got cleaned up as best I could. I had never been on the "floor", so to speak, of a trading company. Clifford met me at the front office with a concerned look on his face saying that he would not be able to spend much time with me.

It looked like all hell was breaking loose. There were about 40 people on the phones, everyone trying to be heard above everyone else. A sense of urgency and desperation were not lacking.

Clifford Morgan was the Executive Vice president for GCAP. Clifford's firm not only sold securities, the firm held positions in many companies as well that they had helped fund during the early stages of development. Clifford mentioned that he had one 40-dollar stock that had plummeted to 15 dollars and things were not looking like they were getting better either. The way things were going he expected it would take years for people regain faith in the market and put money back to work doing new ventures.

Clifford showed me to a room with plenty of space adjoining his office. I set up my keyboard and sound system. Clifford mentioned earlier the he liked Jimmy Hendrix, so I brought along some Hendrix MIDI files to play with. I did a brief demonstration and Clifford tried out the software.

Clifford brought some other people over from the firm, and introduced me to them. That all went well. Clifford told the president of GCAP that I was set up with a keyboard, and some music software. He decided to check it out.

Clifford introduced me to the president of GCAP, Steve Allen.

Steve took a very honest look at Maestro. He liked the automated accompaniment. He thought the software needed a better interface though to show the notes that were being played on the keyboard. That would help people to learn. That was a good idea. I thought I could implement a virtual keyboard, given a couple months of development time.

After the whole demonstration and introductions, I asked Clifford if he thought he could raise money for the project as he had done for Visiosonic. Clifford said the market was different now. Visiosonic raised money during the heyday of the dot COM boom. Now it was much harder if not impossible to fund new projects. I appreciated his honesty.

Clifford had been hoping to take Visiosonic public soon. Even that project did not look promising at this point. His best recommendation was that I contact

Visiosonic and try to work out a deal with them. Clifford said I should contact Joe Vangieri, the president of Visiosonic. Clifford said that everyone called Joe Vangieri, Joe V. Clifford said the Joe V. was a great guy to work with.

At the time I didn't know much about Visiosonic's product line. Clifford told me that it was a dual MP3 player, but that didn't tell me very much. I was more interested in focusing on my own product. I told Clifford that I would keep Visiosonic in mind. I planned to try and get Maestro going on my own marketing it over the internet.

I spent the next couple days looking around New York. I found a couple clubs within walking distance of my hotel in Manhattan and spent some time there Friday evening. I was looking for a karaoke bar. I was told there was one somewhere on the far side of the island. That was a drive though, and I decided to stay within walking distance of the hotel.

I felt a sense of tension in the air in downtown Manhattan. It felt almost like the place was on the verge of some great destruction, though I didn't know exactly what or when I got the feeling that I didn't wanted to be in Manhattan very long. I passed that off as a concern caused by crowds and the traffic, having to be constantly vigilant about avoiding an accident. I was amazed that I did not see a single accident during the trip.

I returned to Varco the following Monday, and caught up on my design work that week. The following week, I put in another vacation request. That request used up all my remaining vacation time, plus 1 day.

I planned to visit Gerald Benjamin at ICR in San Francisco. Gerald was speaking at an event, and he thought that he could introduce me to some investors while I was out there. The event was being held September 12th. I booked a room and bought a ticket to fly out September 11th.

I left for the airport early in the morning on the 11th. Half way to Dallas I heard on the radio that a plane had crashed into one of the twin towers in New York. I had been walking under those towers less than a couple weeks ago. I wondered how someone could be stupid enough to crash into a big building in broad daylight.

Sometime later the news report said that another aircraft had crashed into the second tower. These were full sized commercial aircraft filled with passengers. Listening to the story unfold I learned that a third plane had hit the Pentagon and that a fourth was in the air headed towards Washington being followed by fighter aircraft.

I pulled over to the side of the road. It took a long time for the enormity of it all to sink in. The two towers were burning and collapsing with death tolls being estimated in the thousands. The attack represented a loss of life and destruction equivalent to, if not greater than, Pearl Harbor.

The World Trade Center was the headquarters of many different securities companies. I didn't know whether or not GCAP had an office in the World Trade Center, but I expected they would have many friends working in those buildings.

All flights were cancelled. No one knew the extent of the attack. A somber conversation with Gerald told me that few people if any would be attending his event. I drove back to Austin and withdrew my vacation request.

For the next couple weeks I did not know what to expect. Terrorism on a broad scale was not out of the question. I hated to imagine the damage that could be caused by a small number of evil people bent on causing destruction at any cost.

I decided to buy a gas mask, dosimeter, and iodide tablets. I considered getting a weapons permit, but decided not to unless I saw a real need for it around me. Things seemed to come to a stand still those first couple weeks after the attacks.

Over the next couple months the market crashed. If it was difficult to raise money for private ventures before, it was even harder now. Gerald told me as much. I didn't give up. I recognized that my chances were significantly reduced though.

I continued to work at Varco. I decided to try building up some money to invest in my Maestro project by doing options trading.

My father set us all up with options accounts to do options trading. My father sells covered calls, which is a safe way to trade options. He gets money up front for selling the calls. Usually, the calls close out of the money, and he keeps the money for selling the calls plus the shares that underlie the call option. If on the last day of the option the stock is going to close in the money, my father buys back the options. On the average he makes more money doing this than just holding the stock long and not trading options. This method is a sure method if the stock doesn't go bankrupt. It isn't a way to 10 X ones money quickly though.

To make a long story short I gave my parents a few more gray hairs over the next few months doing very risky options trading. When it was all said and done I lost a few thousand dollars. I figured that I was going to have gray hair if I didn't get out of the public market. I bought a high-resolution video projector to do presentations for investors. I kept the remainder in savings. Predicting the market in the short term is extremely difficult to do, it is close to being random. Long-term predictions can be made, but doing well with options requires guessing exactly what a future stock price is going to be at a precise time. The actual value of the option is based on time, and just the mere passing of time reduces the value of the option unless a major swing happens in the right direction at the right time. Even if someone had inside information, one would still have to know how every shareholder is going to react in the near term to do well with options. Those shareholders all have free will that varies greatly in the short term regardless of long term social trends.

I started working for Varco in the industrial products department. This department was an effort to diversify into sensor projects that were not related to the oil industry. When I arrived at Varco there were nearly 30 people in this new department. Over the past few years, through a series of layoffs, we were down to 2 people in that department. To make a long story short, I found myself unemployed.

I had some savings and some money in a 401K. I decided to try and become self-employed developing and selling music software. To make another long story short, I made a great many improvements to Maestro, but I did not raise the money needed to market it. After the attack on the World Trade Center, and the large dip in the stock market, the job market became very challenging especially in certain technical fields focusing on new product development. I spent quite some time job searching without success.

I decided to give Visiosonic a call. At this point I was interested in selling my music project, or leveraging it somehow to allow me to do something about the whole civilization issue. The whole process of trying to raise money and develop software was taking a long time. As time ticked by learned about the challenges related to starting a new business but I was making little progress getting the information out regarding the physics breakthroughs that people needed to know about. I was not confident that people would take my word for what the future would hold. I expected that I would need to present some products that demonstrated the reality of independent spheres of energy. Putting together enough money to build my own prototype fusion reactor was the first thing that came to mind. Rather than burying the technology in the Department of Defense I wanted to make a much simpler lower cost version available to the public.

I still had Joe V.'s phone number and a recommendation from Clifford Morgan to contact him. I called Joe V. and got through. Joe V. has a New Yawk accent, but he says he is from New Jersey. You can't really be a New Yawker if you're not from somewhere else.

Joe V. said that I called at a good time. His whole development team had just walked out on him and he needed some help.

I was interested in Visiosonic helping me market Maestro rather than doing development work for Visiosonic. I did not rule out the possibility of helping with Visiosonic products. After talking a couple times with Joe V. we decided to meet at an investor presentation that Joe V. was doing in Miami.

Joe V. was planning a trip to Miami in a couple weeks to do a presentation for potential investors at an event hosted by the New Idea Center. The New Idea Center is a group that helps fund early stage projects by matching up qualified investors with entrepreneurs. They also work with smaller established companies. Mick Lopez is the president of the New Idea Center. Mick was a very successful IBM manager. He formed the New Idea Center in Miami, with some other partners, to help improve the economy by getting new projects started.

The event was being held at the Bill Board Live in Miami. The Bill Board Live is a popular club. They own a large three-story building not far from the beach. The Bill Board Live features a dance floor with live broadcasts to MTV, cutting-edge DJ's, and new bands.

Joe V. made arrangements to include me in the event at the Bill Board Live. I arrived at the Bill Board Live early and found the room that the presentation would be in. It was a nice room on the third floor with a large bar built-in bar and comfortable couches all around.

I had never met Joe V. in person. I managed to spot him when he arrived from his description and dress. Joe V. was about 5 feet 11 inches tall, medium build, and a little over 40. Very few people were walking around with suits at the Bill Board Live, so it wasn't difficult guessing it was Joe V., dressed in a suit and carrying a brief case. Another gentleman in a suit accompanied Joe V.

Andy Savas arrived with Joe V. Andy was responsible for raising about half the money to get Visiosonic started, which was about 6 million dollars. Andy was heavy set, medium height, with graying hair and a friendly face.

We all talked briefly before the event. I got to see Visiosonic's product line for the first time, along with a company business plan and financials. I discovered that Visiosonic's PCDJ product was much more than a dual MP3 player. The product was a professional tool for professional DJs. The software automatically calculated the BPM for songs, did beat matching, mixing, cross fading, etc. PCDJ is the market leader for professional DJ software. DJs at the Billboard Live used PCDJ.

Mick Lopez began the event by introducing himself and giving a brief description of the companies that would be presenting. It looked to me like the three companies presenting might outnumber the actual investors attending the event. That turned out to be true. When I say the companies out numbered the potential investors, I don't mean the total people representing the companies. We only had one, possibly two potential investors at the event. The two other companies were also presenting music-related ventures.

Apparently, some other people were expected, but didn't actually make it to the event. Even up to the beginning of 2003 the market had been reeling from the bursting of the dot COM bubble, corporate scandals, the fear of terrorism, and the uncertainty of war.

Joe V. presented last and talked about the Vision of Visiosonic and the PCDJ product line. The president of the Bill Board Live, Mitch, attended the event as well. Joe V. and Mitch got along well and planned to work together in the future doing a remix contest. Winners of the software remix contest would get a recording contract and perform at the Bill Board Live.

After the event Mitch hosted a get together with an open bar and great appetizers. All in all, it was a great event. We enjoyed each other's company, and built alliances with other companies in the music industry.

After the open bar and appetizers, Joe V., Andy Savas, and I went out for dinner at a cafe on the beach front. With all the research I had been doing on monolithic particle physics I was anxious to tell somebody about these things. I presented the basics to Joe V. and Andy. Joe V. said, "What's scary about this whole thing is that I think I actually understand it!" Joe V. is a salesman by nature, not a physics professor.

We got back to talking about music software products. I said that I was considering selling Maestro, and using the money to start on my own energy project. Joe V. and Andy were looking at raising money and doing acquisitions in the future. They thought something could be worked out along those lines in the future. Andy was looking at doing a reverse merger into a public shell. A reverse merger means that a private company buys a public company. The transaction makes the private company stock public.

In the near term, both Andy and Joe V. thought I would be better off working out some kind of agreement based on royalties, or an equity trade, rather than cash. I said I would consider that. We arrange to meet in Clearwater, FL the following week, home of Visiosonic.

Arriving in Clearwater, I met everyone at Visiosonic. I decided I liked working with the people at the company. I moved to Clearwater. Joe V. had a spare room at his house, and offered me a place to stay while I looked for an apartment. This allowed me to get started right away.

We worked out an agreement to add Maestro to the Visiosonic product line. The plan was to break out a "Learn To Play Keyboards" module, and a "Learn

To Play Guitar" module, then market those as separate products using the Visio-sonic branding.

Before I started writing new code I decided to try and help with the Visiosonic fund raising. After being in Clearwater for a few weeks it became clear that Visio-sonic was having financial difficulties.

I decided to try and help with the Visiosonic finances as best I could. I called up Gerald Benjamin at ICR and arrange to visit with him at his office in San Francisco. Gerald was working on putting together events for investors and entre-preneurs around the country, as usual. Talking over the phone we conceived a plan that would involve Mick Lopez hosting an event in Miami, combining the resources of The New Idea Center and ICR. Gerald has a substantial mailing list in the Florida area. He expected to be able to draw a good-sized crowd.

I flew out to San Francisco and met with Gerald. We talked about ideas for the event and I agreed to help organize things for an event in Florida. At the same time, Andy Savas was busy working on the public shell idea for Visiosonic to bring in some cash.

After the trip to California I returned to Clearwater and encouraged Joe V. to do his best to keep the company going. I started working on the Play Guitar and Play Keyboards releases for Visiosonic. I sold rights to my Maestro software to Visiosonic, and the product was repackaged as Play Guitar and Play Keyboards. That covered some near term expenses, but did not result in a future income, and I was still job searching. So, I finally decided that I would need to apply for unemployment benefits.

Neither the public shell idea nor the new funding raising event worked out for Visiosonic. There is a certain time period from January to about March that is a good time to do fund raising in Florida, after that many Floridians migrate north during the hot summer months. We just did not have enough time to get every-thing together within the window. The public shell involved significant up front cash that was just not available.

Joe V. and Andy were interested in helping me pursue an energy project down the road. None of us were in a position to fund a major energy project ourselves. We were not the only ones having trouble with finances. GCAP closed its doors around that time. We decided that a new release of PCDJ would generate cash flow. The current release was more than 2 years old. Visiosonic had a customer base of thousands of professional DJs, most of which would upgrade. I agreed to work on the development project without upfront pay in anticipation for future royalties after the product was released. That allowed me to make some construc-tive use of my time while searching for a regular position or consulting work.

We arranged contracts with two other developers to work on a new release. Sumner McCarty had been working on doing some bug fixes for the current release of PCDJ written in Pascal. I received an Email from Kyle Judd. Kyle is a digital audio programmer whom I had contracted previously to help with Maestro. Kyle, Sumner, and myself began work on PCDJ version 6.0.

The new release of PCDJ took much longer than we all expected. During September of 2003 I received royalties allowing me to become profitably self employed. The royalties were not that much compared to when I was doing consulting work, but it was enough to publish this book. Since the income was a royalty rather than a salary much of my time was also free to promote the book. I published the book through Iuniverse. Iuniverse is a Barnes and Noble subsidiary. They have a great print on demand publishing services for $450.

The information in this book does have value. It might be possible for someone with serious financial resources, which is the one thing I did not have while publishing this book, to use this information to build and patent their own products in the field of energy. Essentially, if that happened, I would lose the opportunity to develop these things myself once a patent was awarded to someone else. Not only that, the individual controlling the patent could do whatever they wanted to with the technology including putting it on a shelf. The clock continued to tick though, and it did not look like it was going to rain money anytime in the foreseeable future, so I decided to publish. My parents were also concerned about me personally, making some of the information in this public. In the end, I decided it was more important to publish everything and see how things went from there. Concerns such as these affected the content of this book. The fusion chapter is based on a real story, but the descriptions of the technology are fictional, and are not likely to help an engineer actually build a fusion energy system or weapon system.

The process of working with people to complete this book manuscript established key relationships. It turned out to be the best course of action writing this book. Energy is not the only advanced technology that can be derived from the monolithic particle model that I have presented. It turns out that life itself is related to very high frequency field of energy. The energy of life has an interesting effect on water. Having said that this entire universe is basically life, I should also mention that diversity is key. Not just any old field of energy is capable of changing the characteristics of water such that the water becomes the very stuff of life itself. I had been playing around with rejuvenation products for some time before September of 2003. With the help of a business partner I decided to make these products available to people.

The next chapter is dedicated to the technology and products related to rejuvenation. In the grand scheme of things, eternal youth and rejuvenation are more important to quality of life than any energy technology. The products that have been developed to date are real, measurable, direct energy products that make use of stable spheres of energy. The energy of these products is enough to rearrange the molecular bounds in water, and produce forms of living water. Life is not a mystery. Life is real, and life including all plants changes the water it comes into contact with by adding spheres of energy to the water that are larger than molecules. A strong living field of energy is able to solidify a mixture of living water and calcium into bone. Bone is more than water molecules mixed with calcium atoms.

Traditional chemistry fails to recognize spheres of energy larger than atoms.

6

Rejuvenation

Many people from history, and many people today, seek solutions related to the problems of aging. Of all these people seeking to regain youth, Ponce De Leon's search for the Fountain of Youth is the most famous. Ponce De Leon found an island off the coast of Florida where not a single native showed signs of aging. The natives could not even remember someone getting "old". The food supply of the island was ideal. The natives drank and bathed in the purest water on the planet. Pools filled with live coral and fresh water existed on the island. Today that coral is dead.

Ponce De Leon visited the island, met the natives, and bathed in the pools. Ponce De Leon did not achieve complete rejuvenation after bathing in these pools for more than a week. So, Ponce De Leon killed all the natives on the island. He left the island to continue on his quest to find the Fountain of Youth. Apparently there was a standing law in Europe ordering the death of all natives found during explorations of the New World.

Ponce De Leon found a fountain of youth. He didn't realize it was a lifestyle, and that the water on the island contributed to longevity. There was a balance on the island between the stresses on the body that contribute to aging, and positive elements of lifestyle contributing to repair of the body. Rejuvenation is actually a much higher goal than eternal youth. The idea of rejuvenation is to tip the scale even farther, and rapidly repair the destructive effects of aging. It is much easier to just not age in the first place. Kind of like the difference between regularly changing a cars oil compared to completely rebuild the entire engine plus doing a frame off restoration.

There must be hundreds of wrong ways to go about trying to achieve health and rejuvenation. I have tried at least a dozen paths myself that haven't worked. In this chapter I will simple present the right path.

Maybe civilizations existed a very long time ago that we know very little about, except stories in the book of Genesis about people living for many hundreds of years. Rumor has it that a long time ago people lived for hundreds of years enjoying vibrant health. Those days are about to return.

The people that Christ cured of various ills still continued to age and die. Even so, the fact that faith healing or contact healing does work at times is a clue towards developing rejuvenation technology. The energy of life itself, when of sufficient purity and strength, is able to lift up other life. This could be compared to magnetic material that has weakened over time as individual atoms within the material go out of alignment. When in contact with a stronger magnet, the weakened magnet can be restored to full strength. That is an over simplification, and it should not be taken as an endorsement of magnet therapy which is more likely to be harmful than helpful. The problem with magnets is that they are stuck in a dipole field mode and magnets are not the kind of energy field that a body thrives on. If you have ever had an uplifting experience being around certain people then you have experienced this do a certain degree.

In modern times 120 years is close to the record for the longest span of life. What we really should be looking at is how long an individual maintains a very high quality of life enjoying vitality and strength. Many athletes achieve that peak in their mid twenties or early thirties.

Breaking a world record for being the oldest person "living" tethered to a bunch of machines is not a desirable existence. Actually, death by old age isn't much different than death by slow torture. The process of healing can also involve some pain, but it is a different kind of pain, transient rather than persistent. Having youth to look forward to at the end of healing does help. Depending on how much rejuvenation is required, the process can take quite some time.

Let us consider the goal of rejuvenation compared to automotive maintenance. Preventative maintenance involves taking good care of a car according to the owner's manual. If there were an owner's manual for the human body preventative maintenance would involve doing all those things required per the owner's manual. People usually do not do that with cars, much less with their own body, so repairs become necessary. I consider a book called "Fit For Life" a good owners manual for the human body. This book was very popular about ten or fifteen years ago. The book is as true today as it was back when it was selling millions of copies.

Using the natural diet talked about in "Fit For Life", and getting some exercise, one can reduce the pace of the aging process. "Fit For life" is at least 200 pages and most of it is essential information regarding how the body works. All I intended to do here is recommend "Fit For Life" as an owner's manual covering preventative maintenance for the human body. In peak condition the body can actually tolerate things like mixing different foods more so than "Fit For Life" implies, but a less than ideal diet can take its toll over time, and most people are not in perfect condition to begin with even during youth.

The goal of rejuvenation technology would be like a complete engine rebuild and frame off restoration. This is a different goal than preventative maintenance. If rejuvenation could be achieved then eternal youth would be easy in comparison. For example, being able to completely reverse 15 years of destructive aging, going from a biological 35 back to a like new 20, one could just rejuvenate every 15 years and break all the preventative maintenance rules in between rejuvenation cycles. In practice, one would just do some of those good things that brought one back to 20 to maintain peak physical condition while stretching the preventative maintenance rules. Rejuvenation does not mean people couldn't be blown up and killed. It is not about regenerating limbs or organs that have been lost. The goal of rejuvenation would be to repair existing systems to like new condition.

Rejuvenation requires very advanced knowledge regarding the exact nature of the body. This detailed knowledge involves concepts that most people do not consider, even in the field of alternative health or faith healing.

The most important aspect of a body is completely invisible. A very high frequency energy field pervades every cell, every organ, and the body itself. These energy fields are the manifestation of life in this universe. The state of those energy field drives the visible physical body if not completely, than more so than anything else. Life is not as mysterious as people think. The technology exists to measure and visualize these energy fields. The subject gets little more than fringe attention though since even if some people recognize these fields exist up until now there has been no practical way of understanding and changing these fields for the better.

Water makes up most of the body. Understanding rejuvenation involves understanding water more than anything else. There is more to water than people have imagined. There is more to the universe than little balls of energy called atoms, there are also bigger balls of energy ranging in size from atoms up to galaxies and every size in between. Balls of energy larger than molecules help bind solids together, rocks and body tissue like bone and skin are good examples. Do you see why skin sags as people age? There is a special form of water that I call "Vital Water". Among other things, the energy of life changes common water into Vital Water. Vital Water contains more energy in spheres larger than individual water molecules.

Water is able to store energy better than many other substances. I mentioned that water could trap electromagnetic spheres. These spheres do pervade all space, but making Vital Water is not as simple as splashing around in the bathtub. A molecule is a group of atoms stuck together. The chemistry of today correctly models each atom giving up a certain number of electrons to a larger sphere of energy to bind the molecule. Taking that one step further, molecules can be bound closer together by even larger spheres of energy.

When pure life energy comes into contact with common water atomic and molecular bonds are changed. To be a little more precise, some oxygen is released and the extra energy held by the hydrogen atom is free to form other bonds. Among other things, this process raises the pH of common water. The water of life is not common water, it is water energized to different levels. It is not as simple as dropping two electrodes into the water. It is necessary to use stable spheres of energy with characteristics that match the frequency of life to create Vital Water. Once again, this is an over simplification, but about as much detail as will present here. Of course, adding other elements to water can also change water to an acidic or alkaline pH. Testing will demonstrate that the products available on the web site raise the pH of distilled water time after time without adding anything to the water besides energy. These products are simply crystals pervaded by a sizable energy field compatible with life, no batteries or outlets required.

Fresh fruits and vegetables are good for you because they contain living water. Cooking, and all methods I know of processing juices, either dilutes or completely drains away Vital Water releasing the extra energy in the water. Garlic contains a more concentrated form of Vital Water inside live garlic cells. Adding more life energy to Vital Water we get something I call Vital Oil. Vital Oil is essential for building bones, the immune system, digestive system, and rejuvenation in general.

Vital Oil is a state of water that involves less oxygen and more energy in spheres larger than atoms. Vital Oil is further concentrated by the body to create solid structures. For example, bones are built by combining some calcium atoms with Vital Oil plus more energy to make the bones solid.

Given a concentrated field of energy at the frequency of life this energy field is capable of changing the material characteristics of water. In the womb an ideal environment exists and the energy of life is concentrated. As energy is added to common water it becomes vital water. As even more life energy is added the viscosity of the water changes. The surface tension of vital water is different than common water. Vital water feels almost like silk. These characteristics help distinguish common water from vital water. The body cannot use common water in the cells or blood stream until it is converted to vital water.

Aging is directly related to the contamination of vital water in the body and the loss of Vital Oil. At some point shortly after adolescence Vital Oil begins to decline and aging begins. Healing slows down as people age, since the body expends Vital Oil in the process of healing. The body ends up having to play a triage shell game, moving Vital Oil from the joints, teeth, or bones to repair damage to critical systems like organs, particularly the digestive system. This becomes a vicious circle. Eventually everything begins breaking down resulting in death by old age. Vitamins make very little difference. The fields of the body become entangled from injury reducing the bodies ability to create Vital Oil and this pure Vital Oil that is largely an endowment from birth is used up over time.

There is very little of this Vital Oil in fresh fruits and vegetables, but it is better than none. Garlic has a higher concentration of Vital Oil but it would have to be eaten fresh or freshly juiced to get it. Fortunately, there is a better way. Extracting Vital Oil from live garlic cells is not an easy thing to do. Garlic sting is very real. In fact, I tested using a drop of fresh garlic oil on the skin to see how far this garlic sting could go, and after it blistered through a couple layers of skin I decided that was a good enough test. Pure Vital Oil can sting a little when applied to an injured area, but the area will then heal faster rather than blistering. Diluting fresh garlic juice and drinking it with other fresh juices can help the digestive system, but it is not an easy thing to drink and there is some trade off since the body has some digesting work to do. Fortunately, there is much better solution that is at least 10 times better than any garlic mix.

It is possible to energize a certain natural crystal material such that a stable sphere of energy capable of transforming water pervades the crystal. Just as an atom never wears out, once a crystal is energized it will continue to transform water over thousands of years. The proof of this that the products work and that any lab can measure the change in the water. A single crystal about 1.5 inches in diameter and .75 inches high can transform 1 gallon of distilled water into Vital Water over a 24 hours period. The crystal will never dissolve or wear out in any way.

Producing Vital Oil requires manufacturing facilities and commercial quantities of different energized crystal material. Larger crystals containing as much as 60 percent gold in a particular crystal matrix are used.

A range of vital water and Vital Oil can exist depending on how much energy is added to the water. Basically, a little energy can be added to the water making one form of vital water, and more energy can be added making vital water that is ready for immediate use inside the cell membrane. Depending on health and vitality, the body has some ability to further concentrate the energy in vital water.

A variety of crystal products are available on the web site to produce forms of vital water. Vital Oil is also available. A higher energy form of Vital Oil called Healing Salve can be made at home. This is accomplished by mixing Healing Salve Mix with vital water. This does not require a commercial facility since the crystals provided in the Healing Salve Mix are being consumed in the process of making the thicker Healing Salve.

The womb is full of vital water and some Vital Oil allowing the baby to absorb pure cellular water and nutrients directly through the skin. Throughout life, the body is still capable of absorbing nutrition directly through the skin given the right conditions.

Using a large quantity of crystal it is possible to convert an entire bath into vital water, or at least make a significant percentage into vital water. Vital Oil can be added to a bath and it will be absorbed directly through the skin. This results in improved skin tone, color, and softness. The change is rapid and the feeling of an entire warm bath of vital water is wonderful. The body will exchange contaminated vital water within the cells in favor of clean vital water from the bath. The skin is a selective membrane, so it won't pull the contaminated water back in after it is sweat out into the bath water. This bath of warm vital water with a little Vital Oil and maybe some Healing Salve added in is one key to rejuvenation.

Going back to a bath of municipal common water that smells in many locations would be difficult to tolerate. Crystal Beads are available that can be used in line with existing plumbing to help purify and raise the energy of water either as it enters the house, or possible in line with a hot water heater. This may require the services of a qualified plumber. The body absorbs water through the skin, not just through the digestive system. Youth has everything to do with cleansing the vital water of the body. In some locations just taking a shower in municipal water could contribute to the aging process. Filters certainly help, but filters do not change the energy state of the water. Water flowing through pipes can actually be stripped of some electrons making it less than neutral in extreme cases.

Vital Oil can be applied directly to the skin where it goes to work quickly healing what needs healing. If used full strength, it may sting. If it does sting that area needs some healing, and that process proceeds rapidly. After applying Vital Oil it can be diluted with vital water if needed. Vital Oil tastes bitter. Adding a little of it to a glass of water, and a little fruit juice for taste, greatly benefits the entire body.

Once upon a time people thought that all illness could be traced to congestion. This is basically true. Detoxification of all internal systems is the first step towards losing unwanted weight and halting further aging damage. Vital water and Vital Oil are essential to any cleansing effort. External signs of aging are almost completely driven by the decline of internal organs, most notably the digestive system. Warm baths of vital water improve the skin rapidly, and can bring some Vital Oil into the body absorbed through the skin. The skin will appropriate that vital water and Vital Oil first. Complete rejuvenation involves rebuilding all internal organs. Things like Grey hair and hair loss are driven by internal decline. Certainly hair can be improved directly to some degree, but it would be an enormous up hill battle without rejuvenating internal organs first.

The products I have mentioned, plus a number of other crystal products, are only available via the book web site. **www.answersforeverything.com**.

The site contains more information about these rejuvenation products, and other health products.

Fortunately, rejuvenation is now possible. It is still a major project that takes many months. Vital water and forms Vital Oil are key, but as I mentioned rejuvenation is a major project and there is more to it. It does take some time and certain dedication. I expect that most people will have the determination to see themselves through a process of rejuvenation. Rejuvenation is about releasing toxins and congestion so that a clean new body can be created. Life is continuously rebuilding the body to some degree. If not, our hands and feet would wear out very quickly.

Rejuvenation Tolerance

Every individual has a unique rejuvenation tolerance that changes over time. Rejuvenation involves processing toxins out of the body. As people age the body's ability to handle and remove toxins decreases. All of the rejuvenation actions to follow need to be taken slowly at first and only increased as the body gains strength and is able to handle faster rejuvenation. The products themselves are not particularly limited in terms of how rapidly they are able to add rejuvenating energy to the body. Rushing out to run a marathon while out of condition can result in broken bones from a fall, torn ligaments, or heart attack. That is the not the right way to get started. Anything can be over done.

A very warm bath of Vital Water with some Vital Oil and or Bath Crystals is a major healing boost for the entire body. Typically, heart rate will increase significantly as the body works to exchange gallons of water and pulls in essential nutrients. Magnesium is an essential element of cell structure and function. Most people are low in magnesium. Vital Oil and Bath Crystals contain colloidal magnesium that is absorbed directly through the skin. No other product can deliver significant amounts of magnesium into the blood stream where it is needed.

There are very few minerals that the body really needs. Colloidal Silver and Gold are useful, but they are not a major essential as is magnesium. Even Calcium is less important. There is so much Calcium put in different things including water that it is virtually impossible to be calcium deficient. The body recycles calcium. For example, as the body tears down the skeleton during aging to extract Vital Oil, the remaining calcium is placed in calcium deposits. During rejuvenation, which the body is trying to accomplish especially during sleep, the body will dissolve these calcium deposits to help with rebuilding bones.

Become aware of physical changes and go at your own pace. The most important part of rebuilding the body is rest and a good nights sleep. If your schedule allows, naps help to. Starting off with vital water baths is the a good way to start and build strength for more rapid rejuvenation.

Advanced rejuvenation

Rejuvenation is about cleaning and rebuilding the body. Fortunately, the body knows exactly what needs to be done and how to do it at a cellular level. You just need to provide a good environment, and the right building blocks, which is mainly pure water at different energy levels and a few essential nutrients.

Rejuvenation includes two key actions. The first is regular baths in warm vital water with a little Vital Oil added to keep the outside of the body in good condition. The second action is periodic internal cleansing. A range of internal cleansing actions exist from drinking a little vital water to doing a raw fruit fast. A little Healing Salve and or Vital Oil can be added to a glass of water to rapidly rebuild and detoxify the system.

The first action involves a good quantity of crystal to reform a whole bath of water. Fortunately the crystals never wear out. A bath is best with some Vital Oil. A shower converter is available which converts water rapidly as it flows around a crystal matrix under pressure.

Most people associate clear soft skin with youth. Just as you can paint over a car with a bad engine, so can the skin be cosmetic.

Internal cleansing is the most challenging one part of rejuvenation. The amount of Vital Oil added to some vital water to detoxify the system would vary for each individual. Start with the smallest amount necessary to produce an effect then build up as tolerance increases. Depending on how much destructive aging needs to be repaired, this could take many months. Rather than doing this all at once I would expect most people do so in stages.

A little Healing Salve added to a glass of vital water is almost tasteless. As the digestive system absorbs all this pure water and healing strength the body will put it all to work. The body has the potential and the innate knowledge necessary to fix and rebuild just about anything. Lets take a closer look at the nature of solids, since the body is essentially in the business of repairing and building solid structures. The body is even capable of breaking down and removing virtually any undesirable solid structure in the body via the immune system.

Just as the term vital water represents a range of waters pervaded by different energy levels, so Vital Oil is used as a term to represent thicker forms of water with even more energy stored in energy fields that are larger than individual molecules. A whole new field of science would be needed to properly describe this; it could be called the chemistry of solids. It is important to gain a basic understanding of this. The body contains many solids, including bone, which is a solid concentration of Vital Oil binding calcium together.

For hundreds of years people in the mining industry have noticed a sort of Saint Elmo's fire and other odd electromagnetic phenomena while crushing large rocks, especially gold ore containing quartz. All rocks contain at least a little if not a lot of energy, in spheres larger than molecules. This energy binds grains of material together. All of the solid crystals on the book web site are essentially man made rocks pervaded by a significant amount of energy. You could say that the planet is basically a rock pervaded by a large sphere of energy called the magnetosphere. You may notice some of the Crystal Beads seeming to "glue" together after drying. There is no glue; this is purely an energy effect. The pyramids were built using precise molds. Sand was poured into the mold and then a solidifying agent was poured in to turn the sand into rock. This is not advanced technology, it is just technology that was lost.

Healing Salve Mix is a combination of dissolving crystals and crystal powder. It takes about 24 hours to make and it can be mixed at home. The result is a Healing Salve that is thicker than Vital Oil. Healing Salve will heal anything that I know of internally and externally, including different types of cancer. Cancer is not "healed" per say. Healing Salve is simply boosting the immune system to a level at which it is capable of destroying all the cancer tissue. There is more to cancer. Cancer is a major disruption of the body's energy fields, which is why people often go into remission even after the cancer is completely removed. Fortunately, there is a solution.

Cancer is an example of cellular desperation. Cells are alive and sentient just like everything else in this universe to a greater or lesser degree. When the environment of the body becomes to toxic, or the energy fields of the body become significantly disrupted for whatever reason, the cells stop working for the greater benefit of the body. Basically, the cells go out of control. The body has trillions of cells, and the immune system is constantly dealing with at least a few cells here and there that have become cancerous. So, cancer is not a black and white issue, it is a matter of how much and to what degree the body's police force keeps things under control.

This chapter is not intended to diagnose or cure any particular case. The information is presented for review, and the actual accomplishment of healing is left in the hands of the individual. Invasive medical solutions will never cure cancer. Drugs will never cure cancer. Until people understand what cancer is, change life style, and give the body what it needs, individuals will just get more cancer after doctors remove the immediate threat to life. That is if doctors are able to remove the immediate threat cancer poses without ending life in the process. These medical solutions are emergency solutions at best, a more painful drawn out death at worst.

The reason why the medical community considers cancer impossible to cure is that even after removing the cancer the disruption in the energy field of the body still exists. It might take that field disruption some time to build another sizeable cancer, but it will eventually unless the body is truly healed at a spiritual level. Aging is both physical and spiritual in nature.

A cancer diagnosis should not surprise anyone. It is simply a label placed on a condition that has existed for a very long time, usually decades. One would have to be extremely out of touch with their own body not to recognize such a non-optimum condition existing for such a long period of time. More likely people know something is wrong, but do not go to a doctor because of a lack of confidence in the medical community. The medical community is doing their best with what they know.

Looking at society today, many people in their mid twenties or thirties need to start taking action against cancer. Even if it isn't diagnosed as such, aging and cancer (benign or not) go hand in hand. A pot belly is usually an indication of abnormal growths in the digestive tract. Given a perfect digestive system a body does not get significantly over weight. Whatever the cause, correcting a less than perfect digestive system is the most important part of a rejuvenation cycle. Every part of the body is related to the digestive system. In most cases fixing the digestive system will handle whatever other problems exist which are usually secondary problems.

The first thing an individual facing cancer must do is have no fear. Emotions do change the physical state of the body, creating an environment conducive to either healing or further decay. The emotion called fear accelerates and contributes to the cellular problem we call cancer by adversely affecting those energy fields which define the body. A sense of serenity, action, or artistic creation are all positive frames of mind that do help.

There are many good reasons why one should have no fear dealing with cancer. First, maybe the cancer will encourage a life style change and the cancer "victim" will end up healing themselves and understand the importance of a holistic healthy life style. Second, the physical body is such a transient thing anyway that fear of death is just not something one should be concerned about. A bomb, bullet, or any number of other things in ones immediate vicinity can take away the body very quickly. There is no way to guarantee that won't happen at any time in this universe, so why fear death? It does make sense to be concerned about spiritual existence. Bombs and bullets cannot touch the spirit, nor can cancer.

The individual spirit can exist in a very good state or a very bad state after death, and for a time span approaching eternity. Those different states can include all the emotions of life, including rapture and pain. Actually, the range of emotions and sensations that can be experienced spiritually is greater than what the body can experience. So, the spirit is the important thing. The third reason for having no fear is that the whole thing is solved and the solutions are in this book.

The individual needs to choose a course of action to regain health. That is part of the will to live. In extreme cases the invasive medical procedures of today may be necessary. A holistic approach needs to be used in any case to achieve true health and vitality, even if combined with an invasive solution. If an invasive solution is selected, it makes sense to strengthen the body for some weeks or months first to endure the trauma. Considering how long it takes to develop cancer in most cases a few weeks isn't going to make much difference anyway. Some people might change their mind and realize that an alternative holistic solution will work.

The products I mention prove themselves. There is nothing in the products except water and some minerals as far traditional science goes. Therefore, these products will be broadly available as a natural holistic solution.

Here is how the Healing Salve Mix works. The contents are poured into a container and mixed with warm vital water. Shake vigorously. You will notice many bubbles. Oxygen is being released from the water creating a thicker energy rich form of water called Healing Salve. Let sit for about 6 hours, and repeat several more times. The Healing Salve should web between your fingers. Externally, apply to any array of concern. It may sting for a little while. Rinse as necessary. A little Healing Salve can be mixed with a little Vital Oil in a glass of vital water. One teaspoon of Healing Salve and a half-teaspoon of Vital Oil would be a good start. Adjust the amount down or up as necessary in that same ratio. Healing Salve mixes perfectly and immediately back into water, as does Vital Oil. This little drink, which is almost tasteless, will clean and rebuild the entire digestive system over time taken a number of times per day.

When you feel like resting, rest, when you are hungry, eat. Hunger will diminish while the immune system goes to work and the body is busy cleansing and healing.

Would it make sense to try and rebuild the engine of a car while the engine is running? You could try, but you would have a very difficult time doing so. Completely repairing the digestive system is the key. Repair of all the other organs that are supporting the digestive system will fall into place automatically. That is why fasting does work. A fast gives the body a chance to repair things. A 30 or 40 day fast is difficult to do, but if the body has enough Vital Oil in reserve it can accomplish a great deal during a fast. Fortunately, a true fast is not necessary. Raw fruit is absorbed high in the digestive tract with little effort. That is better than fasting on nothing, since the body needs energy to rebuild. The quantity of fresh fruit should not be limited. At the same time, adding anything to the diet other than fresh fruit will throw a wrench into the rejuvenation process possibly stopping forward progress completely. Remember, achieving rejuvenation is no small thing, the body is always subject to stress and healing quicker than the time can degrade things is a real challenge.

Given plenty of Vital Oil and Healing Salve for use in the immune system and tissue repair the body will do a good job of cleaning things up without forcing the body to consume good tissue. Even with a starvation fast the body will consume bad tissue first. Fresh vegetables can also be added while rejuvenating, as long as the vegetables are not eaten with the fruit and you wait 2 hours to digest the vegetables before having more fruit. The stomach has to work a little harder to digest fresh raw vegetables, but they are absorbed high in the digestive track if they are not cooked. Cooking vegetables or processing fruit turns them into toxins to some degree putting stress on the entire digestive tract. This is especially true for fruit. Fruit really needs to be fresh not processed. Lightly steaming vegetables is ok, but they are best fresh.

The only thing I have found that comes close to the cleansing effect of Vital Oil is 3 or 4 ounces of freshly juiced garlic mixed with collard green juice. The garlic juice is almost impossibly to drink though, resulting in sweating and an immune system response directed at live garlic cells in the juice. What is happening is that the live garlic is very strong so the body has a battle digesting the fresh garlic juice. Even so, this has a net positive effect if one can endure it. Some Healing Salve and Vital Oil well diluted with Vital Water is much more effective even than the garlic juice, and tastes fine. Add a little fresh fruit juice and it tastes very good. No amount of fruit juice will mask garlic juice.

Digestive cleansing does involve frequent and less than solid bowel movements, plus bloating. Bloating is often part of the healing process as the immune system goes to work. It does take months to repair decades of poor diet and less than optimum health. Understanding that this is a necessary part of the healing process prevents people from giving up and not recognizing progress. A flat stomach, and an ability to consume cleansing fruits and vegetables plus energized water without any result other than great health and vitality, is the end goal. Anyone who looks old has a digestive system that needs major repair. Every part of the body regenerates and is capable of healing over time. The digestive system cells, especially the inner wall of the digestive track, regenerate most frequently.

A fresh fruit and raw or lightly steamed vegetable diet is necessary to clean and repair the digestive system. There is no limit to the amount of fresh fruit and vegetables that can be consumed during cleansing. Fruits and vegetables are the best food on the planet.

Besides dealing with material toxins, and providing the body with the energy it needs to heal, digestive cleansing is further complicated by the presence of parasites. This is another rude topic. The reality is that anyone enjoying a typical U.S. diet has parasites, especially if an effective cleansing program has not been recently completed. Parasites are life forms that have adapted to evade the body's immune system and survive in the environment of the digestive system. Some herbal products are able to chip away at parasites and kill them without harming the digestive system. Effective herbal products are available on the book web site. It can take 90 days or more of herbal bombardment, combined with a cleansing diet, to completely deal with parasites.

In addition to all the points above, it is possible to straighten out the flow lines of larger energy fields pervading the body, essentially recharging the system. If the magnetosphere were at full strength, rather than 10 percent of what it was 2000 years ago, just walking around bare foot would help. Walking bare foot at the beach still helps a little.

A variety of crystal blankets and pillows are available. These products provide a very good nights rest, and do help revitalize the living energy that pervades the body. Larger solid crystals provide an even greater energy boost, and allowing healing energy to be concentrate at different locations.

Larger solid crystals are available one the web site that are pervaded by a sphere of energy that has a very beneficial effect on the energy fields of the body. The best of these products is called a Crystal Stack. The Crystal Stack is shaped like a circular pyramid.

One day I will build, or show someone how to build, a 3D visualization device to show the energy fields that create the body changing in real time. These fields are very real and very measurable. Just as the magnetosphere is invisible, yet moves a compass, so are the fields of the body invisible yet direct the growth of cells into a beautiful or not so beautiful body. These energy fields are much more important that the actual visible structure, a live energy field actually creates physical structure. Aging and cancer are two common examples of major problems directly related to a decline in strength or a disruption of the energy fields that pervade the body. If you have every noticed an "electrified" feeling being around certain people, or touching certain people, this is an example of one persons energy field improving the state of another persons. The feeling can be mutual. It would be very interesting to show people these fields in 3D using some technology, but that project is a ways down on my list because it is not essential at this time. The energy fields pervading the body, and changes in that field, can be experienced easily. The ability to simply relax and perceive the body is an essential part of rejuvenation and healing.

Fortunately, cancer cells are qualitatively different and can be identified by the immune system. The body's lymph system is responsible for keeping the body clean of foreign material, bacteria, viruses, and any cells or tissue that have become cancerous. The lymph system contains 2 or 3 times more fluid than the entire circulatory system. Essentially, the tissue of the body is suspended in lymph fluid, and the lymph fluid helps transmit nutrients from the blood and remove wastes. The lymph system circulates by muscular contraction. The lymph system does not have a heart pump like the circulatory system. That is why a sedentary life style promotes toxic build up and slows down healing. It does not take extreme exercise to get the lymph system operating. Brisk walks, deep breathing, and massage are all very helpful. Even 30 minutes of slow walking helps.

Some people have had a degree of success against cancer through fasting. In fact, I would say that anyone who has really cured himself or herself of cancer has done so via a holistic solution that included fasting. A fast on vegetable and fruit juices is usually best. The reason why a fast helps is because the body frees up energy that was being used for digestion and puts that energy into the immune system to consume the bodies own tissue. That is a very good thing. The body is very smart about burning the bad stuff first during a fast. The reason why people fail trying to fast is because hunger can become unbearable when the body does not have enough white blood cells on hand to speed up the immune system. When the immune system is very active hunger reduces to tolerable levels. Making enough white blood cells requires vital water and Vital Oil concentrated as available energy in the bone marrow.

The first thing that a person notices starting a fast is usually "stomach growling". This sensation is more accurately described as an active lymphatic system. The lymph system surrounds the stomach as it does all the major organs. This growling is actually a good thing. Using the Crystal Stack, as I will describe, one should occasionally notice a sensation of lymphatic motion, or "growling", immediately after touching different locations with the Crystal Stack. That just means the body is healing rapidly. The body will not commit to tearing down a congested area if the energy is not on hand to do the job. The energy disruption called cancer impedes the body's immune system. Using these rejuvenation products a major immune system response and healing becomes possible for individuals who would not otherwise have the energy to clear up and heal sizeable areas of the body.

Recognizing the true nature of a problem is the first step to solving it. The converse is also true. If one fails to solve a problem, one has not recognized the true nature of the problem, and must go back to the beginning and try again. Only then can an honest search for the truth begin. Criticizing another person's ideas, no matter how apparently misguided, is a sure indication that an individual is not honestly searching. That is a clue to help recognize the individual leaning towards evil. An honest critique, presented with understanding and following logically from another's ideas, actually helps the search for truth. These two things should not be confused.

Being created in the image of God is not about having long gray hair and resembling some fantasy sitting on a throne in space. Being created in the image of God is about being a spirit that transcends death and influences this universe via energy fields that pervade an area much like God pervades the entire universe. Mankind is important because the spirit of man is more like a fallen angel than the spirit of an animal. An animal has a spirit, and that spirit is focused on the cellular details of building the body. The planet could be said to have a spirit focused on building the planet. The spirit of man does not build the body directly. The body is built by another spirit similar to that of any animal. The spirit of man should be focused on helping manage creation, and bringing beauty and love into the universe. Keeping the body in good condition shouldn't require hardly any attention at all. The spirit of man could just as easily rise above the body and help out around the universe at will, which happens all the time looking at guardian angels and people inspired by the holy spirit. Love is an experience of attraction and understanding between people. This is a very real attraction, and the spiritual equivalent of gravity.

Without the technology to see things directly, how can a person become more aware of the energy of life pervading the body so as to fix it if necessary? As the body goes through life cells are going to get damaged, this cannot be avoided. Cellular damage can disrupt the fields pervading individual cells and this disruption can be passed down to new cells when the cell divides. Even if everything is in place in terms of nutrition, and every molecular requirement met, these old cellular injuries can cause problems in new cells. That is why people have had trouble achieving rejuvenation. The only solution is a spiritual solution dealing directly with the life of each cell.

Larger spheres of energy that are compatible with the life of the body and individual cells can straighten out these energy disruptions at the cellular level, at the level of organs, even energy disruptions driving cancer can be fixed. That is why faith healing and the laying on of hands sometimes works. An individual who is very strong spiritually can help straighten out a body, though this would be limited to some degree by the tolerance of the individual doing the faith healing. Healing a bunch of people at the same time would be more difficult than one, and the amount of healing needed would also influence things.

The important thing to realize is that this is a universe of life, even an atom could be consider to have some hint of awareness and free will demonstrated by quantum mechanics. As cells release the pain of old injuries one will experience a brief flash of pain in that area of the body, or possibly just a tingling sensation. All the rejuvenation products I mentioned helped accomplish this to some degree. These tingling sensations and brief pain flashes make one aware of changes in the energy fields of the body.

The Crystal Stack available on the web site is pervaded by a strong energy field capable of restoring clean energy flows through the entire body. Using two or three Crystal Stacks is best to allow for flows between Crystal Stacks releasing more disruptions that are blocking the normal flow of energy that pervades the body. This will work to some degree regardless of an individual's state of mind, but it works even better in a relaxed state of mind simply perceiving the body and perceiving the changes driven by the energy of the Crystal Stack. Essentially, the field pervading these Crystals could be consider pure new life with great potential but little or none of the knowledge and experience that is part of the spirit of men and women that allows people to do things like build civilizations. Life is continually created into this universe by heavenly bodies. I use the term life loosely since I view this entire universe as life. Certainly some things are different than others. The Crystal Stack is pervaded by energy from that 90 percent of the universe that was once called "dark matter" just because people couldn't find it. It is not "dark matter", it is just invisible life energy that is everywhere. These invisible spheres of energy are not nearly as dense as atoms, so they are invisible.

The Crystal Stacks are the key to rejuvenation, providing a tool to strengthen and straighten out the energy fields that pervade the body. Applying the concept of rejuvenation tolerance, one would simply relax at some distance from the Crystal Stacks rather than holding them to begin with. Do not be surprised if touching a Crystal Stack to a particular location produces what feels like a physical shock. It just means that there is a fair amount of energy tied up at that location from previous injuries. Simply approach this area gradually and eventually it will clear up leaving little more than a light tingling sensation.

The following sequence of actions will produce rapid results no matter what the problem is. If the results are to rapid simply reduce the amounts, times, or distances used for each step. Add 1 once of Healing Salve, and ½ once of Vital Oil to a glass of vital water and drink. Some fresh fruit juice can be added for taste if one likes. Rub some Healing Salve on any area of concern. This should leave a little Healing Salve on the hands. Lie back and relax on a Crystal Pillow and grasp a Crystal Stack in each hand. Relaxing music can help. Simply relax and percieve any change or lack of change throughout the body. It may take a few minutes for the crystal energy and the energy of the body to find common ground. Try touching different parts of the body with the Crystal Stack.

This action will repair all physical and spiritual aspects of the body, done at regular intervals over time, until particular rejuvenation goals are achieved. Please understand that the spirit of man and women is not the spirit of the body, nor the energy field of the body that I am talking about regarding rejuvenation. The spirit I am talking about fixing up is directly related to building the body and is no different than the spirit of any animal. Advancing the true spirit of men and women would result in individuals like Jedi Knights. That is not going to happen using a few of these Crystal Stacks. One day I may write another book along those lines.

For comparison, different religions on this planet claim to directly benefit the spirit. I do believe that a few extraordinary Buddhists in history have done some amazing things. However, Buddhism does not provide consistent results. The cancer rate of advanced Scientologists appears to be higher, or at least equal to that of the general public. I know some of these Scientologists. If Scientology was spiritually beneficial it would benefit all life of a spiritual nature. That is certainly not happening uniformly within Scientology organizations. My personal opinion is that Scientology can have negative spiritual results as often as positive results. Sort of like mankind playing with atomic bombs without knowing all about physics, Scientology seeks to improve the spirit without knowing enough about the spirit and life in general. If Scientology management had a clue regarding this universe they would understand the physical challenges that this planet faces. I would not like to think that they did know and decided not to tell anyone the true nature of things. That would be treason against all mankind, no matter what justification was presented to explain keeping things confidential. Having spent many years studying Scientology, and counting many Scientologists as friends, I believe it is important to make this point. I am quite sure that people are wasting their lives in Scientology waiting for confidential levels to be released which will certainly not help address the near term physical challenges that will have to be faced on this planet. The spiritual challenges will be greater. In my opinion Scientology does not offer a solution in that spiritual respect, given the current situation.

Back to using the Crystal Stacks. You learn by experimenting. Try moving the crystals around and holding them differently to see what affect that has. Energy is centered on the central axis of the Crystal, so the orientation is significant. Try touching different parts of the body to the Crystal Stack. You should notice some tingling, or brief little sparks of pain. This represents negative energy being released that has accumulated in the body. Larger areas of injury may release a little more energy. What is happening is that this energy once tied up by old injuries is released back into the main field of the body where it can do some good. In the case of major problems like cancer, the immune system will be activated to destroy all the cells affected by the cancer. This would result in discomfort, or an ill feeling, that will pass in time when the immune system is done its work. Cancer is not something that is going to be fixed over night. Depending on the size of the area affected, the body may need to virtually rebuild the entire digestive system for example. That would require regularly repeating this process of adding energy to the body over a period of months, and doing all the other good things I mentioned diet and rest wise to complete the healing process. Making rapid healing progress is important dealing with cancer. A cancer has become part of the body and can grow much faster than other tissue. That means it has to be destroyed by the immune system faster than it can grow.

It is important that the body have Vital Oil and plenty of vital water available. The immune system is the body's secondary digestive system. When the immune system is very active the sensation of hunger reduces or goes away depending on how busy the immune system is. The immune system white blood cells essentially eat other cells, bacteria, viruses, etc. The by-products return to the system as nutrients or toxins to be flushed out of the system.

Considering all these things about energy and the body, how can it all be summarized in terms of energy driving motion of the body at a cellular level up to athletic events? In brief, the body burns hydrogen for fuel. It is the hydrogen in carbohydrates, not the carbon, which releases energy when combed with the oxygen that we breathe. Carbon is simply a waste product exhaled as carbon dioxide. The energy of life concentrates water, releasing some oxygen from the water, and freeing some hydrogen to be used as energy. Water can change and become as solid as bone as more energy goes into spheres larger than molecules binding various molecules and atoms into solids. Driving all this body work is the spiritual nature of cells and the body itself. Pure living water at various energy levels, and products that concentrate an energy field compatible with life, can be used to strengthen and purify the body and life itself.

Rejuvenation tolerance summary

The previous rejuvenation steps describe methods of maximizing the speed of rejuvenation. Beginning a rejuvenation program an individual should use less of everything and see what affect that has. For example, pure vital water will probably be to much to start with never mind adding some Healing Salve Mix. Try mixing a little vital water with crystal water made using the Crystal Beads first.

Removing toxins

The environment today introduces many different toxins into the system. One example is aluminum added to many different products like deodorant. Rather than trying to list out all the negative products that contain toxins, or trying to avoid toxins, I will focus on methods of removing toxins from the system. Toxins are anything that doesn't belong in the body. Toxins slow down the healing process.

A product called Crystal Mask is available on the web site. Crystal Mask is mixed with a little water to form a paste. As the paste dries it will pull toxins out of the body from as deep as a half inch under the skin.

A water energy system has also been developed which creates a large negative field in water to pull toxins directly out of the body in a bath or foot spa. This system consists of an electronics package combined with a spherical array that is submerged in the water. Using vital water the detoxification process proceeds at a rapid rate. There is nothing solid about atoms. Electromagnetic forces hold atoms together. Free radicals and toxins that do not belong in the body are almost always positively charged. The body is trying to get rid of these things as it is. The water energy system provides an attractive force compatible with the frequencies of the body. Water provides a pathway to remove many of these toxins.

Free atoms are not very solid and move about quite easily. Atoms are not visible of course. Toxins are pulled to the center of the orb through the water and then group together and exit the top of the array. The body is not inclined to reabsorb these toxins once they have been neutralized and clumped together. Significant amounts of toxins are removed, enough to change the color of the water and float on top of the water.

The water energy system does not make vital water. Vital water is created at the atomic level using solid crystals. The crystal products already mentioned are sufficient for achieving complete rejuvenation. If the water energy system being tested proves to be a significant benefit it will become available on the web site.

Common health misconceptions

Many false conclusions are promoted regarding health. No matter how much advertising is spent on Coca Cola, Coke as it is degrades health compared to a refreshing glass of water that is essential to life. If Coke was made with vital water it could be beneficial providing the water of life plus carbohydrates. Today, that is not the case. Coke is acidic and stored in a toxic aluminum can that leaches aluminum into the acidic water. Just one of the dozens, or hundreds of things in the environment, all converging on the health problems we see all around us. It would be a mistake to blame the Coca Cola company. There will always be companies that provided whatever people are willing to pay for. The correct focus would be public education. Once people are educated they may still decide to destroy their own body. There isn't anything that can be done about that. People do have freedom and it would be a far worse thing to try and force something personal on someone.

Taking a bunch of manufactured vitamin pills is not the solution to rejuvenation or most people would be rejuvenated by now. Vitamins are a distant second compared to the other things I have talked about. Absorbable vitamins are found in fresh fruits and vegetables. Absorbable is the key word. Dried raw food supplements from a combination of things like wheat grass, alfalfa, certain types of algae, seaweed, etc. do provide a great source of natural vitamins and essential nutrients if prepared properly. Essential oils cold pressed from natural sources are just as important, if not more important than vitamins. I would continue any vitamin program that was producing a noticeable benefit. Personally, I see no reason why massive quantities of vitamins well above the USRDA should be helpful, nor have I experienced that to be so. To be on the safe side, I use a tablespoon per day of concentrated vitamins and nutrients from natural sources contained in a product called Vital Nutrition Plus.

Cooking sends molecules flying all over the place releasing larger energy sphere. Sometimes these larger spheres settle back into water, which is why soup broth can by very good. That is why a fresh carrot is crisp and a cooked carrot is mushy. Old skin sags and young skin is firm. Once again, there is more to life and solids than the molecules that today's chemists understand only partially. Raw fruits and vegetables are better than cooked. Though I believe people guess this to be the case, maybe knowing why will change people's habits.

The body does not need nearly as much protein as people think. Mother's milk is only 1 percent protein. The body does not use protein directly. The body has to break down protein, and hope that it gets the right amino acids to build up human protein. These amino acids are all present in a pure ready to use form in raw fruits and vegetables. It takes a great deal of energy for the body to digest meat, and additional energy to deal with the higher level of toxins found in meat compared to fruits and vegetables.

Some people have used this energy equation to "trick" the body into burning fat. I would submit that Dr. Atkins looked his age, and is now dead, as evidence that his program is no better health wise than any other path. It is true that the body will burn fat, being starved of carbohydrates and having to deal with getting all that protein and fat out of the system. There is some risk that the body will concentrate some toxins in a hard to burn off layer of fat lacking the energy and cleansing power to neutralize and properly eliminate those toxins. Fruits and vegetables help cleanse the system. The Atkin's diet plan avoids fruits. Fruits are nature's perfect food. The only saving grace of the Atkin's diet is that it is not as bad as the typical meat and potatoes plus desert at the same meal diet plan. The stomach has difficulty digesting meat and starches at the same meal. Following that up with simply carbohydrates that should go directly to the small intestines makes things worse. This can result in undigested stuff leaving the stomach to ferment and putrefy in the intestines. This is one cause of rapid destructive aging and a relatively early painful death by old age. So, there are worse things than the Atkin's diet. Overall, the Atkin's plan it is not a very healthy solution. With a little dedication, people will find that far more fat and toxic tissue can be burned on a diet of fresh fruits and vegetables. This diet plan will continue to burn the weight without negative consequences until a trim, firm, energetic body is achieved.

During the prime of youth, when all body systems are operating at peak condition, the body can tolerate most anything diet wise for some number of years. Eventually poor diet catches up with people in the absence of an ongoing rejuvenation program. Once the body is rejuvenated, continuing the non-diet elements of a rejuvenation program can offset the decline that would normally be caused by poor diet and life style habits.

The sun does not cause cancer or degrade the condition of the skin. In fact, getting some sun actually helps destroy cancer and improve the skin. Ideally, the body should be able to tolerate a fair amount of time in the sun without getting sun burned. This tolerance does vary with genetics. The body needs some sunlight to feel good and produce certain vitamins. Moderation is the key. A bath of vital water with some Healing Salve mix and Bath Crystals will help up a sun burn quickly restoring the skin to a condition better than it was in prior to the sun burn resulting in a higher tolerance for future sun bathing. This is true for minor sunburn, not a major burn resulting in blisters for example. The same bath would also help heal a major sunburn but restoring the skin to like new condition will involve more than that. The Crystal Stack product would be needed to relieve cellular memory of major damage that can persist after the visible damage has healed over.

Vegetarians do not suffer from protein deficiency. Philosophically, I am not a vegetarian. I do enjoy barbecue, pizza, etc. At the same time, I recognize that a raw fruit and vegetable diet is the best thing for fat burning, and helping to rebuild the body. Doing this right and completely once, many years of youth can be gained during which time is not necessary to do another rigorous rejuvenation cycle.

Is there a limit to how many rejuvenation cycles one can go through, or some kind of diminishing returns over time? The answer is no. It is a misconception that aging is inevitable or that time itself has a direct effect on anything. It is true that damage can accumulate if it is not repaired. It is also true that damage can be repaired, and that the body can grow positively in the face of a challenging environment. Besides losing parts of the body over time that cannot be rejuvenated, the body is capable of healing anything. For example, the rejuvenation process covered in this chapter does not grow back teeth that have been removed, but existing teeth structure can be greatly strengthened. In theory, there is no reason why technology could not be developed to grow a new tooth using cells from the patients own body. This would be an extremely advanced technology though since there is more to a tooth than a few cells. An energy field pervades each tooth and defines the structure of that tooth.

Dairy products have become popular, and like Coca Cola this is one habit that has broad negative impacts. Nature did not intend for humans to chase down cows and suck on a cow's udder. Cows have a stomach that is very different than a human stomach. Cow's milk contains casein, which is used in industrial glues, and cannot be digested by the human stomach. Consumption of dairy products degrades health. Butter, which is the fat portion of the milk, can be digested. I do like cereal. I drink Rice Dream, which is one of the best tasting beverages I know of besides water, and is certainly better than milk on cereal.

Fat Burning

Fruit is the ideal food, but it does have to be eaten by itself. The stomach uses different enzymes to digest different things. Mixing things at a meal makes digestion more difficult; exactly what food combinations are better than others is covered in "Fit For Life". Fruit doesn't need digestion in the stomach; fruit just gets absorbed in the upper intestines. Fruit before a regular meal is ok. Fruit after the meal creates a problem. After waking up, it is best to only consume fruit for few hours. The body goes through natural cycles of acquiring food, assimilation, and elimination.

What would you say about someone who bought a car but didn't know anything about changing the oil? What if you saw that same person pouring Hershey syrup into the oil system and then throwing a few pop rocks into the radiator? If that person were planning on entering their car in a race against you next week you might sell him more syrup to pour into the oil system and tell him that the radiator was suppose to sound like that. Or, if you earned a living fixing cars, you might not encourage people to read the preventative maintenance section of the owner's manual.

Considering the prevalence of unhealthy foods and life styles in society, and the false information mixed with a little truth in the field of health, it is not surprising that so many Americans do not enjoy perfect health and vitality. One person at a time is the only way to change that.

Rejuvenation is a lifestyle, not a quick fix. The environment and the life style people lead can contribute to aging. The question is, can one repair damage to the body quicker than damage occurs? Reversing 10 years of aging in 6 months would be quite an accomplishment. It is certainly worth doing.

How about looking like the people on the cover of Muscle And Fitness? Obviously, regaining youth is going to help achieve that goal. For older people, it would be wise to focus on detoxification and rejuvenation first then work on strength training or body sculpting latter. The reason is that hard exercise tears down the body so it can rebuild stronger. The body has priorities in terms of what it is going to rebuild and repair first. Repairing internal organs is more important than strength building. The visible signs of aging are driven by internal decay. Adding the stress of tearing down muscles via hard exercise to the task of rejuvenating the body can be done, but it is not a wise thing to do. Just a little light exercise is best to get the system moving during a detoxification program.

Looking like the people on the cover of Muscle And Fitness is a very challenging goal. Extremely low body fat is probably not desirable in the grand scheme of things but it is cool to achieve it once, take a picture, and then go back to a more comfortable body fat percentage. I have found that the fastest approach to losing that much fat is a diet of raw fruits and vegetables appearing on a list of "negative calorie" foods. Some fruits and vegetables take more calories to digest and assimilate than they provide. That means eating more of these things will burn more fat than a fast on just water would. A complete list can be found on the net using the key words "negative calorie". The list includes broccoli, carrots, pineapple, grapefruit, mangos, and apples. Servings are not limited, and it helps to have 8 or 9 servings of these things during a day. That is the fastest fat loss plan I know of.

The first casualty of any battle is the plan, and "the battle of the bulge" is no exception. I suppose the better the plan is going in, the better the plan's chances of not being totally abandoned. This negative calorie diet with exercise is a good plan. Even so, things may arise that shake resolve or sow doubt.

Net weight can be misleading. The goal is to increase lean body mass, which is denser than fat, and reduce fat. Guys gain muscle quicker, and tend to burn calories quicker, than girls. Theoretically, doing bodybuilding, a guy could put on 20 lbs. of muscle and lose 20 lbs. of fat, looking perfectly ripped, without a change in net weight. That is an extreme example just to make a point. Some exercise is needed to make any fitness plan work and to burn fat. Exercise gets the blood moving, delivering the various hormones and things required to make the fat burn. All that said, this negative calorie raw food plan with exercise should result in a consistent and considerable drop in net weight, even while building strength.

It is important to drink as much high quality water as possible. Colder water burns a little more calories. A 16-oz glass of water weighs 1 lb. The water does not add fat. So, realize that net weight can fluctuate a few lbs. in a day, up or down, even if one is actually burning fat that day. Food and water have weight. It takes at least a day, if not two days, to completely process things. There is a delay moving energy into the blood stream.

The body will strive to keep a relatively constant energized blood stream, whether it gets the energy from positive food calories, from fat, or stores extra energy as fat.

Metabolism determines the rate at which energy is pulled out of the blood stream. It is much easier to burn 1000 calories by optimizing metabolism than it is to run off 1000 calories. A pound of fat is 3,500 calories. One thousand calories is getting close to what an athlete burns doing half of a marathon. Marathons are for people already in optimum condition. The value of a 30-minute aerobic session is not so much the calories burned during training session, but the fact that the exercise session raises base metabolism for some hours after training and gets the blood moving. When people are out of shape carrying extra weight risk of injury during hard training is highest. The wise thing to do is to listen to one's body and train, but don't train through pain. Ideally, exercise should be enjoyable and free of pain. If it is not realize that some healing needs to be done first.

Doing this negative calorie diet, I would do the program for some number of days, maybe 4 days or maybe 10 days, and then I would break off it for a day or two. That helps prevent the body from adjusting by dropping base metabolism. It is also easier to break the larger program up into smaller goals. Things like Metabolife, or Xenadrine RF-1, do help speed up metabolism. These products should be used in moderation if they are used at all. I have found that creatine increase strength and endurance rapidly without any adverse effects.

Drugs

Vibrant health, friends, loved ones, and fellowship provide an unmatched joy and thrill of living. Illegal drugs, and even some legal drugs, destroy general health. Many false promises are made in the name of drugs. Drugs can be used as a tool to manipulate people via addiction for monetary or sexual gain.

So, why are so many people including many professionals using drugs?

Considering the prior lack of rejuvenation technology in society, and the generally high level of toxins in the environment and in processed food, I would say most people rarely experience truly vibrant and perfect health. Drugs do mask the pain temporarily. Maybe that is why some people turn to drugs.

Maybe so many people have fallen into the trap of drugs that these large groups of drug users just drag everyone else down with them in and unending vicious circle.

Maybe some people have never experienced true spiritual freedom. Drugs deal such a blow to the body that the spirit can begin to involuntarily separate from the body as the body approaches death in a way. Sort of like a near death experience. Whether that experience is good or bad is beside the point, the experience is temporary, involuntary, and quality of life degrades over time using drugs. Drug use over time reduces the chances of an individual ever achieving true spiritual freedom, particularly at this time in history. There are exact spiritual reasons why that is. For the moment people will just have to take my word that drugs are bad from both a physical and spiritual point of view.

Maybe some people are followers rather than leaders. A real follower will do whatever they are told is "cool", even if that involves driving nails through their own head without anesthesia. The only way to change such people would be to change the viewpoint of the majority. If the vast majority begins to recognize drugs for what they are things might change.

Or, maybe some people consider themselves "rebels" and automatically do anything that is discouraged by the majority. It is not easy to determine exactly why particular individuals turn to drugs. The reason I just mentioned though probably have something to do with it.

Drug use is a very personal thing. Drugs cannot be legislated away. As a society we recognize the destructive effects of drugs and try to intervene and legislate against it at a personal level. It would be nice if simply writing on some paper in congress could create personal enlightenment, courage, common sense, or any number of desirable personal characteristics. Life just does not work that way.

The only solution I see is to change one person at a time. There will always be rebels without a cause who decide to go against social ideals no matter how positive, logical, or enjoyable those ideals are. A single person like a big brother or a big sister can help someone avoid a life of destruction. A group of positive friends can help pull other people out of a life of drugs. It is not possible to legislate away a personal problem like drug use. We can legislate against harm caused to others, thereby limiting the negative social impact of certain individuals.

Even though I do not expect legislation to eliminate personal drug use, I would not change existing legislation involving street drugs. There is a great need for better people at this time in history. As a society, we need to make some effort to steer people away from self-destructive habits like mind-altering drugs. Hopefully there will be enough people who choose to create a better civilization rather than becoming trapped in the illusions of drugs.

Beyond personal rejuvenation

Having knowledge and ideas for a better future, and knowing that other people share those views, but not knowing how to reach them, is a difficult position to be in. Knowing is one thing. Having products to demonstrate that knowledge is another thing. How many people would believe that rejuvenation was possible if the means to rejuvenate people was not available? How many people would believe that there was anything to be concerned about along the lines of global catastrophic change if they didn't see the environment changing?

Knowing that rejuvenation is possible, and a very long life of youth and vitality awaits, people should be more interested in preparing for the future. It would be very disappointing to spend a year or more actually regaining ones youth only to die a handful of years later in a natural catastrophe along with the rest of mankind. Even there was only a low probability of this catastrophe materializing, it would be a good idea to spend even as much as a decade making sure that one was physically prepared for any worst case scenario. In my opinion the odds are about the same as the odds of an apple hitting the ground after it drops from a tree. That opinion includes a certain sense of destiny that is not completely quantifiable. It may be that honest calculations could be done that include the physics models I have presented which give better odds extending the timing of earth's next magnetic reversal.

It will not be possible for individuals or groups to prepare for the events I describe without public support. The advanced technology that will be required will not be made available to the public until priorities are changed at a national level and national security laws and policies are reviewed and changed by the executive branch. That just isn't going to happen without 150 million people in the U.S. demanding change. Publicly supported preparations can only begin once a majority elects new officials and laws are changed. A democratic nation acts with consent of the majority.

How long does it take to reach and educate 150 million people? However long it might take, that needs to be done first. If the individuals who read this book do not consider these things important enough to get personally involved in political campaign then change is just not going to happen. That is just the stark truth of things. Personally, I have done as much as I can writing and publishing this book. I cannot knock on 150 million doors and try to explain that there really is a good reason to read and understand this book beyond the cool cover art.

The knowledge in this book does threaten multi billion dollar special interest groups dealing with oil, health care, drugs, centralized energy distribution, etc. For that reason alone it will be difficult at first making things like new energy products available to the public. The faster these ideas become popular, and the more people take advantage of these rejuvenation products, the better the chances are for success. Knowledge is the important thing. If 1 million people really understand these models and products it would be very difficult for a special interest group to eliminate a small threat and put everything on a shelf. In time, any bright engineer could go down the same path developing similar products for the public.

During 2002 I was appointed to President Bush's economic advisory council. I was invited to attend events in Washington. I did not do anything with this appointment initially, and I suspect it was just a fund raising gimmick, but I decided to get involved anyway. It makes more sense getting involved in politics when one has a good reason for doing so and a well defined plan that one wants to see implemented. I have been as guilty as anyone in the past regarding politics, rarely voting, though I have provided moral support for the Republican Party talking up some Republican candidates.

I decided to run for president. I turned 35 in April of 2003, making me old enough to run for president 2004. However, the race in 2004 does not leave much time for campaigning.

In politics taking a stance on an issue can involve alienating people on the other side of the issue. Deciding to run on the Republican ticket for example would alienate people in the Democratic Party to some degree. That is something I would like to avoid. My real goal, and the only chance I believe this nation has, is working together.

The Republican Party is not perfect by any means. Given a choice, there are specific reasons why I would choose the Republican Party over the Democratic Party if I had to choose a major party. There are good reasons why a president has never been elected outside a major party.

Our two major parties have over one hundred of years of history behind them. That is why most people are affiliated with one of the two major parties. In a democracy ruled by the majority it takes a majority to win. That may sound so obvious it isn't worth mentioning. Party loyalty does exist for good reasons, and that loyalty has been built up for generations.

7

The Republican Party

I believe that many Democrats misunderstand the Republican Party, thinking it is a party for the rich or special interest groups. That is just not the history of the Republican Party, nor does it reflect that attitude of most Republicans.

People with military experience usually back the Republican Party. One reason for that is because the Republican Party has always supported defense. I believe the Democratic Party grossly under estimates the nation's need for defense in the not too distant future.

On a personal level, Republicans tend to gravitate towards God and Country more so than the average Democrat does. I am just talking about things in general, people I have met, and what I have heard from leaders in different parties. I am not saying that Democrats do not faith in God and Country. Certainly they do, more so I suspect than the average person in most other nations. I just see this as a stronger value in the Republican Party.

I consider the history of the two major parties just as important as where the parties stand today. Going back 150 years, not too many generations have passed between the founders of the major parties and the people of today. Family values are passed on from generation to generation. Party values are retained over the years as well.

The Republican Party was formed during the 1830's. During those days the nation was torn apart by a moral debate regarding slavery. The Republican Party would decide that debate, once and for all, for the future of the nation.

John C. Freemont was the first Republican Party candidate for President. John Freemont began his career as a lieutenant in the Army's Topographical Corps, which later became the Corps of Engineers. Freemont was given the mission to map routes west, through the mountains, into California, helping achieve Manifest Destiny. Manifest Destiny was the idea that the United States would expand from sea to shining sea. Freemont's maps helped settlers go West. Settlers used Freemont's maps to find the best land to settle on and avoid dangers making the trip.

Freemont is not well known today. In those days he was very well known for opening the West. With little more than a platoon of frontiersmen, including Kit Carson, Freemont secured the entire state of California at the beginning of the war with Mexico. Freemont was a man of determination and principle.

The American settlers in California joined forces, with the help of Freemont, rather than being driven out of California by Mexico. Freemont was brilliant on a field of battle. He naturally commanded the respect of his men. Freemont commanded several successful expeditions out West. Buchanan, the Secretary of State at the time, briefed Freemont on a special mission. Buchanan said this mission had the support of President Polk, which was probably true. Freemont was asked to make sure that California did not fall into the hands of Mexico. We were not actually at war with Mexico at the time, and California was not yet even a territory. Buchanan said that Freemont would have the support of the Navy if hostilities were to arise.

Freemont led his men through a treacherous mountain crossing in the middle of winter. Miraculously, his team made it, without the loss of a single man. Freemont's expedition arrived at Sutter's fort toward the end of winter. His men recovered from the journey and picked up supplies.

Freemont met many of the local American settlers. These settlers were concerned about being driven out of California. They wanted Freemont to lead them. Freemont did not know if war had actually been declared, or even what the general situation was. In those days the only way to get a message to California was by ship often sailing all the way to the bottom of South America and back around to California. The panama canal did exist, but it was a long journey in any case.

Freemont would not officially take command of the settlers in California. However, Freemont advised the settlers, and they did unite. Rather than being thrown out of California, the American settlers found themselves in command of California. A "bear flag" was created, and that flag flew over California for a few years.

When the naval commander of the theatre arrived on the scene, announcing that war had been declared, he found that California was already secured. The commander appointed Freemont governor of the territory of California. The battle was already won in California so the navy was able to move on to other missions.

General Kearny of the U.S. Army was given orders to deploy to California. Kearny was anything but brilliant in a tactical sense. Kearny attacked an area in California that did not need attacking. He lost 30 percent of his forces and was wounded himself before Freemont arrived to pull him out of the mess. Freemont even recovered a number of artillery pieces that Kearny had lost during that engagement. Freemont was now a Colonel, with verbal orders from both the Navy and the secretary of state to take charge of California. General Kearny had no written orders regarding governance of the territory.

The navy admiral who put Freemont in charge of California was at least of equal rank compared to Kearny. Holding the rank of Colonel, Freemont was the second ranking officer in California at this point. Kearny demanded that Freemont turn over his position as governor. Freemont asked to see some orders to that effect first.

Freemont knew that the territory of California would not be in good hands with Kearny. Kearny tricked Freemont, ordering him to ride back east with his command then arrested him on the way for a court martial. News of the trial spread quickly. Freemont would not disclose his dealings with the Secretary of State and the President. Technically, when Freemont arrived in California, the U.S. was not actually at war with Mexico and those verbal orders could have caused some problems had they been made public. Freemont would not risk releasing the information, even if it would help him avoid the court martial. He hoped the president would come forward on his behalf.

In the end, Freemont lost the court-martial. He was dismissed from active duty. Shortly after that the President pardoned Freemont and gave him back his commission. Freemont was a little upset that the President would say nothing publicly. Freemont gained some satisfaction knowing that he had accomplished his mission. At the same time, he did not like politics, and did not want to work for Kearny, so he decided not to reenter military service at that time.

The slavery debate began to heat up after the war with Mexico. After the court-martial Freemont used his entire savings from military service to buy some ocean front property in California. Freemont had just served an entire year as governor of the California territory. During that tenure Freemont would accept no personal benefits from his position, which he was offered on more than one occasion.

The land Freemont wanted was to be purchased on Freemont's behalf by an agent. Freemont was still on the East Coast near Washington after being court-martialed. He wanted the land to be secured will he made the trip out west doing a privately funded mapping expedition on the way. He wife would arrive by ship sailing through the Panama Canal. Freemont was upset when he discovered the agent had not purchased the land he wanted, but rather some different land. The agent did actually try to get the land Freemont wanted, but the title was not clear to it.

A few months later gold was found at Sutter's mill. Sutter was a good friend of Freemont. The two spent some time together after Freemont's expedition marched out of the mountains to re-supply at Sutter's Fort. Sutter had recently built a mill on a river. Digging into the river to build the mill, Sutter's men struck gold. Freemont had inspected the rivers and terrain during his mapping mission, and thought that he had found fools gold in some of the rivers.

Arriving on his new land, Freemont discovered millions of dollars worth of gold in the riverbeds alone, plus 20 deep veins of gold on his large new tract of land. A million dollars was worth more in the 1850's than it is today. Freemont looked at his land closely. He had 100,000 acres. He hired a geologist who helped map out the 20 different veins of gold on his property. Freemont offered people 50 percent of the gold that they panned out of the river. That put many people to work for him quickly, and filled up half a building with gold. His wife was pleasantly surprised after enduring a very difficult through Panama and out to California.

Things where going well for Freemont. Freemont and his wife Jessie toured Europe for almost a year in style.

A few years later, southern Democrats were looking for a presidential candidate. The main concern of Democrats in the South at that time was preserving the slavery status quo. The southern Democrats argued that they were bringing the savages out of Africa, and giving them Christianity. Many Northerners argued that all men were created equal, and that included everyone, without an exemption clause based on country of origin, or color.

The Democrats knew Freemont was popular in the North where the Democratic Party lacked strength. Freemont's father in law was senator Benton, a leading democratic senator from the South. Three Democratic Party leaders met with Freemont offering him the Democratic nomination. They viewed Freemont as having the potential of uniting the nation and avoiding armed conflict over slavery. The condition was that Freemont would have to support the existing slavery act.

Back then the major parties did not use primary elections. At that time the only other major parties were the Whigs and the Native American party, the Republican Party did not exist. The Whigs were declining. There can be little doubt Freemont would have won the presidency if he had accepted the Democrat's offer. Freemont was very popular in the North. People knew Freemont had personally mapped the trails West, and appreciated his help and military experience. Many people followed Freemont's court martial in the papers, and even though Freemont lost people liked him and recognized that a certain injustice had been done. They knew that he was responsible for bringing California into the union, which was a huge boon to the nation. California was rich in fertile land, and gold. That gold added to the treasury played no small role improving the global standing of our new nation.

Freemont spoke with his wife before responding to the Democrats offer. Freemont's wife Jessie did want to live in the White House. At the same time, Freemont was completely against slavery. Even though the slave act was legal legislation, Freemont knew he could not speak in favor of it. Morally, he could not accept the offer of the Democratic Party, even though he really did want to be president. Jessie knew that would be his decision when she heard the terms of the Democratic Party offer, and said she would be happy with whatever decision Freemont made.

Freemont respectfully declined the Democratic Party offer the next day. The Democrats were furious. They told Freemont in no uncertain terms that his attitude would lead the entire nation towards civil war.

Other influential people in the North, Seward for example, shared Freemont's viewpoint. They decided to form a new party called the Republican Party. Freemont would be their first candidate. Freemont and Jessie went from abandoning the White House to being part of a new party all in a matter of weeks. Jessie helped manage Freemont's campaign. At the beginning of the campaign support for Freemont and his wife Jessie grew rapidly. They both played an active role in the campaign.

During those days people traveled around and jumped up on stumps to talk up the party candidate. There was very little mass communication. Candidates seldom spoke on their own behalf. Abraham Lincoln was a young lawyer "stumping" for Freemont. Freemont recognized that Lincoln had great potential listening to him one day.

The Democrats nominated Buchanan to run against Freemont.

Freemont was looking like he had a good chance of beating Buchanan halfway through the race. Then, the Democrats launched what may have been their dirtiest slander campaign the Democratic Party has ever run. They printed massive numbers of leaflets, calling Freemont a bastard child. They said Freemont used slaves in his gold mines. In truth, Freemont split gold 50 / 50 with people panning on his land and paid the best wages in California. How many people actually knew the truth? The Democrats said Freemont was a cannibal, and drove people without mercy to death in the mountains.

During one of Freemont's expeditions his group was trapped in the mountains during a very hard winter, they could not make it through and were forced to turn back. It was looking like most of the people in the expedition were not going to be able to make it all the way back to the last outpost where supplies would be available. This outpost was hundreds of miles away through blizzard conditions. The outpost was the home of Kit Carson. Kit had been part of several expeditions led by Freemont, but had decided not to go on this particular one. Freemont sent a small group of the strongest people out hoping they could move faster and bring a rescue team out to save the rest expedition while they conserved their strength. Freemont remained to preserve order at the camp. It may be that Freemont still hoped that with re-supply they could continue the expedition. When rescue did not arrive as expected Freemont struck out on his own to find the team sent ahead. The rest of the expedition began making their way out of the mountains as best they could. Freemont found the advance team half way back to the outpost. They had given up and killed the leader of their team, who Freemont was sure would not have given up on them. Freemont picked the rest of them up and got them back to the outpost. Kit Carson lead out a rescue team with fresh horses and supplies saving the remaining members of the expedition who had not already died from the cold and starvation.

The negative campaign being run by the Democrats began to have an effect. The Democrats cried, "A vote for Freemont is a vote for war!" Actually, Freemont had a plan to move the South away from slavery gradually, using federal funds. It is easier to communicate a negative than it is to present a detailed plan to the public.

Support for Freemont began slipping. Freemont's Republican friends insisted that they respond in kind and slander Buchanan. Freemont said, "You will say nothing!" Freemont would not debase himself by sinking to such low tactics.

Freemont lost to Buchanan 1.3 million votes to 1.8 million. The Republican Party was born. Freemont would not risk dividing the Republican Party in the next race. Many people wanted Freemont to run again, others thought it would be best to go with a new candidate. Freemont threw his support behind Lincoln. Lincoln went on to defeat Buchanan in the next race. Shortly after that the South seceded and began attacking northern positions. Lincoln was not an experienced military leader and the Southern Democrats viewed him as weak.

It is true that Lincoln did not have military experience. Even after the South started attacking the North, Lincoln delayed. Lincoln did not bring sufficient force to bear quickly enough to prevent a long drawn out war. Initially, the South captured many northern armories, making the war much more difficult. For many months Lincoln hoped the South would return to the union even as they destroyed Northern forces and captured many military resources.

Finally, Lincoln decided that he would have to commit and decided to commission Freemont as a major general. The North was short on equipment initially. Freemont personally bought 10,000 stand of rifles, plus artillery, and answered Lincoln's call in Washington. Freemont used his own money and lines of credit from banks in Europe backed by his gold mining operations. When the equipment arrived, Lincoln decided to reroute Freemont's equipment to General Scot, leaving Freemont in a tough position commanding Missouri. In California, Freemont's house was taken over, and converted to an artillery battery. The house overlooked the San Francisco bay. Freemont was never compensated for the loss of his home.

Freemont freed the slaves in Missouri sooner than Lincoln would have liked. Lincoln fired Freemont for this. Grant, who was trained by Freemont, went on to win the war for the North. Freemont died living in an apartment, surviving his wife whom he loved dearly. He had been forced to sell all of his interests, and during his final years he relied on income from his wife's writings published in journals and newspapers. Many of these works survive to this day and were used to tell the story of Freemont's life in a book called "Dream West". "Dream West" tells a tale of adventure from the early days of the United States. There wasn't much during those days that Freemont wasn't involved in.

More Americans died in the civil war than the total casualties of all the wars fought since, combined. Over 200,000 young men from a total population of roughly 8 million died in our civil war. After such enormous losses it is a miracle that the British, the French, and the Spanish did not take advantage of a nation weakened by civil war. That is probably because not to many years prior to the civil war the U.S. defeated both the British and Spain.

Two brothers never fought a tougher fight than the North versus the South during the American Civil War. The North, led by Lincoln and the Republican Party, fought to free all people. It was not just about slavery. We also fought to free the South from the guilt they were piling on their hearts through the institution of slavery.

Declaring any man or women less than human opens the door to acting the same towards anyone. Freemont saw this tendency in the Democratic political campaign against himself. He wrote about this in his memoirs.

Today, many of the ancestors of those people the Republican Party fought and died for, align themselves with the Democratic Party. Personally, I believe debts of that magnitude span generations.

The Republican Party has a history of standing for equality and freedom. It is true that some members of the Republican Party are financially well off. Success requires hard work and honest dedication more than anything else. I am sure that destiny also plays a role in things.

8

Economics

Today we have a free market economy that uses paper as a unit of exchange. We call this paper money. To make things easier we represent paper money electronically in the form of debit or credit cards.

At first glance it looks bad having a paper money system without an equivalent value of tangible property in Fort Knox to back up the money. Actually, the existing system is the ideal system for a number of reasons, even without an equivalent value in gold to back the paper money. The important thing is to avoid counterfeit money. Fraudulent money destroys the economic system.

Consider a 100 percent increase in Gross National Product (GNP). GNP represents private goods and services that people want to exchange. Doubling GNP means that we need twice as much money to represent all the new stuff that people may want to exchange.

If the federal government had to back up all new money with gold or other tangibles, how would we do that? We would have to take an equivalent value of tangible products out of the private sector and put that in government storage like Fort Knox. Doing that would tie up resources that might otherwise be used by people.

Real value exists in useable products, companies that produce useable products, and the labor of people with professional skills. Paper money is nothing more than a symbol used to exchange real products. Gold has value because it can be used by industry to create products that require a good conductor. Gold also has value because people like gold and wear it as jewelry. The gold in Fort Knox needs to stay in Fort Knox since it is a tangible public resource kept for future generations much like our national forests are preserved for future generations. At $350 per once the 261 million ounces of gold in Fort Knox is worth roughly 91 billion dollars. Eventually, someone will figure out how to run lead in a fusion reactor and create gold. Right now that technology does not exist. It is good for a nation to have tangible stockpiles in case of emergency. Currently, the U.S. has very little in the way of tangible stockpiles besides gold. Gold stockpiles should be preserved, maybe even increased, along with creating other stockpiles that would be needed in an emergency. Adding to the gold in Fort Knox would help increase the market value of Gold thus increasing the total real value of the rest of the nation's gold. Basing a monetary system on a stockpile of tangibles is not a rational monetary system when the realities of economic expansion are taken into consideration, but some stockpiles are still a good idea.

GNP represents the real value of a nation's currency. GNP represents the value of goods and services that are being created by the nation, and can be bought using national currency. If too much money is printed, or too little, inflation or deflation results. The law of supply and demand will take its course. It is true that we can influence economic activity a little by adjusting interest rates. That is just one piece of the puzzle, sort of like one man trying to lift a truck off the ground. That one man can try, but he will need the rest of the football team to make it happen. When it gets right down to it people need to get busy, invest, and start producing more stuff to really change the economy.

If most people are earning more money, and inflation is low, that means we can all buy more stuff. Essentially, that is because we are all making more stuff. I call this the "more stuff" economic theory. All of economics boils down to how much stuff, what kind of stuff, and distribution of the stuff. Money is desirable because it is a basic unit of exchange that can be converted into any product.

The reason for all this focus on stuff is that a whole lot of new products will be needed in a short period of time. These new products will include high tech survival systems, new transportation, new energy systems, etc.

If we are going to need more stuff we should understand exactly what a physical product is. Every product that we produce now, or will produce in the future, is made of at most two things. The first thing is natural resources that we pull from the ground. Energy falls into the category of natural resources regardless of how we get the energy. The second thing is labor. Service products are completely labor.

People will know how well the nation is doing economically by measuring the results of labor and how well natural resources are leveraged. Technology that releases an enormous amount of clean low cost energy would greatly improve the economy. Many other things take care of themselves. Fortunately, the U.S. is rich in natural resources. We have vast public lands that are open to any citizen for prospecting. The U.S. has great lands that have already been prospected and yield great resources.

A communist would argue that people managing capital are doing no labor. In most cases that is false. How many people are employed by newly financed ventures? In a few cases this is true and some people sit back doing nothing with a bunch of money stuffed in a mattress. It is not worth taking away the fruits of everyone's labor and giving everything to a government because an ignorant or lazy person might run into some capital. Most fools are quickly parted from their money anyway.

It does make sense to take a close look at how people and companies are gaining and managing capital. If someone is not making a social contribution, things like consumer boycotts could make a difference. More important than boycotts, one could support projects to crowd out less desirable ventures. For example, moving the public to fusion and hydrogen as an energy source is better for everyone paying the oil industry. At this point we have used most of the oil on U.S. soil. The cost of importing does affect the economy. If the balance of trade tips to far in the wrong direction the quality of American life could be affected..

In an expanding free market economy paychecks increase as competition for people with valuable skill increases. There is one thing that can defeat rapid growth. What if a whole bunch of people became scrooges and piled up all the money under a mattress refusing to spend it, invest it, or work on creating new products? The money sits idle, and no progress is made. That would be bad. The parable of the servants, each entrusted with certain funds according to their ability, comes to mind. How much is not so important as how well one does with what one has from a philosophic point of view. From a practical point of view, hoarding tens of billions is going to slow down the economy more than hoarding 1 million.

If the money sits idle, and the money represents goods and services, chances are that new projects are not going to happen. People could all sit around saying they can't find a job and saying they cannot afford to buy or build all the new products.

People with greater wealth should be taking a greater share of responsibility. I would not be surprised if support is withdrawn from individuals who fail to administer that responsibility wisely. One might say, "I have no wealth, so don't look at me." That is not true. Everyone has the potential to accomplish great things. Granted, some are a little better at it than others. The roughly 280 million people in the nation, out of 290 million people, who don't have great wealth do have much more when they work as a team. With 200 million people deciding they want to do something, kicking in a dollar or so each, they will get just about any project done. An exception would be a nuclear sub, which would cost about 20 dollars each. Guess who is really building up the defense department? It is the average individual, paying average taxes, along side the individuals working in defense related industries.

Consider a company socking away 50 billion dollars giving nothing back to society. Say this company makes no significant investments percentage wise, compared to how much they are socking away. Say this company withholds paying even a dime in dividends to the shareholders of the company, who might otherwise invest that dividend return elsewhere. Unfortunately, this has been the rule rather than the exception in corporate America lately.

Things are improving. I was talking about Microsoft of course. I noticed Microsoft did start paying a dividend and doing some community based IT investing right before this book hit the press.

I am not discouraging putting money in the bank. That is necessary at times to build up enough for certain investments. Leaving it there, not investing it, not growing the economy, these things I am discouraging.

A reduction in GNP, or a failure to produce the right products over the next decade, could result in the loss of many millions of lives that could otherwise have been saved. Individuals may not be able to see the stark truth in this until that day arrives. Still, that day will arrive, as surely as the sunrises and sets.

If GNP, and producing the right products, is so important should the free market economy be changed? It should not be changed. A free market economy is one of the basic rights and principles that make this nation what it is. National principles have endured for generations. Hundreds of thousands of American men and women have died defending those principles. A free market economy gets so much more accomplished than communistic slavery that it would be foolish to consider anything else. People are willing to work on important projects producing high demand products. That is not the problem. The problem today is poor leadership that does not start those projects, or even actively apposes new projects.

Until an individual truly understands the value of basic rights, and is willing to put their own life on the line to create a better society, that individual will be subject to manipulation by any group that decides not to play fair and is willing to pay for dirty deeds done dirt cheap. A truly free people is a group of individuals that cannot be swayed by physical threats. After one revolution, and two world wars, the U.S. has proven to be willing to pay the price of freedom at a national level. It is always easier to go along with the status quo than it is to try and change things. It is always safer to keep a low profile, or an average profile, in any movement that seeks to change an existing power structure. When the stakes are very high, an all out charge into action is the best way to go. That charge should be a well planned charge without fear not a foolish charge. There is a difference between courage and reckless action. Courage is taking on a difficult task fully aware of all the risks involved. Reckless action is just doing something for the sake of doing something.

The path of spiritual decline begins with a fear of death. It goes down through a fear of pain, fear of verbal abuse, and then a fear of embarrassment. Eventually an individual gets so bad off they can't even walk out of their own home and look directly in the eye of a fellow American.

In the eyes of God, looking down on one nation under God, courage and holding to principle in the face of any challenge are admirable things. The U.S. is the greatest nation on earth at this time in history. Probably even throughout history. I know that many philosophers of the past would be proud, including the Greeks who gave us democracy.

The solution to economics is three fold. First, social consumerism will help focus the actions of corporations in a broad sense through votes with the dollar. Certainly this has to be balanced with high quality products and fair pricing. We can have our cake and eat it to. Second, investors are needed who are willing to invest in projects that will help secure the future of civilization. Third, political activism is needed, electing better leaders and removing barriers to expansion. Political leaders at all levels who will help better manage resources in the public interest. Political leaders with a plan, not people focused on some narrow interest. It is fine to work towards bringing business to one's state. The important thing is expansion in general.

The Bush administration is trying to stimulate the economy. President Bush is using a dividend tax cut as his economic centerpiece. As it stood before the tax cut, corporate profits were taxed at 60 percent when passed on to investors. The question is it best to have the government manage most of the money or private citizens. Granted, private citizens could be doing a better job right now and investing more in new projects. The new tax bill reduced that 60 percent to about 45 percent. I believe that will help the economy. What I would really like to see is the government contributing more to private ventures on a cost-sharing basis. It is in the national interest to boom the economy by getting new energy technologies out into society. I believe that the federal government is preventing that without good enough reason through the actions of the executive branch and the department of defense. It may be that elected officials are left in the dark as far as that goes. Even so, it is the responsibility of an elected official to find out and do right by the nation.

Eventually, I would like to see a tax incentives related to personal and corporate expansion. For example, if an individual manages to increase their personal income during a particular year, that person should be rewarded with a tax break for that year. I am not saying we need to change existing tax brackets. This would be a bonus just on that year of expansion to reward and acknowledge those individuals and companies that are doing a good job. If a company posts expansion for the year and creates new jobs that would be a good reason to make dividends for that company tax exempt.

Doubling GNP in a short period of time, say over a 3-year period, is a very doable thing. We haven't had a real sense of urgency as a nation since World War II. Much of that doubling of output could be efficiency gains created by automation and better technology. Certainly, much of the increase would be putting more people to work. The urgency of the near term situation merits putting in longer days when necessary. During World War II this nation put forth an enormous effort and displayed a sense of teamwork and patriotism unmatched in history. We new German concentration camps waited if we failed. Granted, the threat is not so obvious as it was back then.

I am sure that some people who are currently retired possess valuable skill sets and abilities. Bringing some of these people back into the work force would increase GNP. Many of the tools we use today, like advanced CAD / CAM technology, computer driven automation, etc. have progressed while people have been retired. Even so, the basics of mechanical engineering and electrical engineering have not changed much. Certainly the skills needed to manage people have not changed.

If seniors make good progress using rejuvenation products I would hope that some seniors decide to enter the work force again. The way Social Security is structured this would not reduce benefits in many cases. An individual could always go back to being retired if they wanted to.

A clock-punching attitude can be detrimental in many work environments. An individual producing a great design or a great product in a ten-hour workweek is doing far better than someone dragging out the same project through two 60-hour weeks. Granted, there are some jobs like running a cash register that are directly related to time.

If all of these private sector initiatives do not result in a labor market that exceeds full employment, then something may need to be done with public money. The reality is that far more work needs to be done in a short period of time than there are people available to do it. This will true no matter how efficient people are at getting things done. Increasing efficiency is absolutely necessary if there is to be any chance of building a self sufficient civilization in about 10 years. Essentially I am talking about building systems that are owned by individuals and families. These systems would provide all the basic essentials like energy, shelter, pure water, oxygen, food, and transportation requiring very little maintenance. The key to all these things is energy. That key does exist today. It is just a matter of enough people realizing that a solution does exist, and making sure that certain anonymous individuals who like the power status quo as it is do not prevent the release of new technology.

A public solution would involve a public investment in additional factories to produce the products that I have mentioned. People would be employed by these public ventures earning money to buy the products that they need. Does this sound like just to obvious a solution to employment problems? Do you wonder why no one has ever tried this? Publicly owned utilities do exist, so this solution would not be without precedent. Could it be that someone is pulling the strings in the private sector keeping the unemployment rate at a particular level to control salaries? I am not going to try and prove that, I am just going to try and set things right. Salaries will increase naturally to reflect increases in GNP and a greater availability of products. There is no sane reason to promote unemployment. That may not even be happening, I just wanted to make that point regarding employment and salaries clear.

Ultimately, it is most important that people be given the opportunity and resources needed to create and own decentralized systems providing for their own basic needs. It would always be possible to sell such factories back to the private sector if congress decided that is what they wanted to do at some time in the future.

The economy is what we all make of it. Energy is so important to economic growth that it deserves special attention. Energy leverages physical labor. Today nations pull most of their energy out of the ground in the form of fossil fuels. Even fission reactors rely on uranium, and uranium is a limited resource at the rate it is being used up by older reactors. Newer breeder reactors change that equation. The days of non-renewable energy are coming to an end. How many acres could a farmer farm if that farmer had to go back to hand held tools or oxen pulled plows? Not enough acres to feed many people. The same is true in many industries. That is why energy is such a driving force in the economy.

9

Energy

The global oil situation is closer to critical than most people are led to believe. We have been led to believe, or many people assume, that we have 100 years of oil in the ground around the world. This is not the case. The exact situation regarding oil is not something that is covered much in the media. The power of media groups is not so much what is said, but what is withheld from public awareness. Why would the media mislead us regarding oil reserves? The goal of the oil industry is to pump all the profitable oil out of the ground and sell it at the highest possible price. If people knew how critical the situation was, and the economic catastrophe that would certainly follow the beginning of real shortages, then people would demand new energy sources now rather than pumping every drop of oil at the highest possible price.

Special interest groups are not necessarily positive or negative, people are entitled to a difference in opinion. It is a mistake to label anyone or any group purely good or evil. That is a mistake because doing that ones fails to truly understand the group or individual. Even if a group is bent on a path that will result in the demise of everyone, it is still important to take a very calm objective look at how that group influences society and come up with a very sound tactical plan to implement change. There is a difference between a relatively ignorant terrorist who would like nothing more than to blow himself up in the name of Allah, and a group that is working towards a particular goal like world domination. It might be that high up in the oil industry people know exactly what this planet will face regarding a near term pole shift, and that group chooses to build up their own stock piles of various resources seeing no solution to save any significant percentage of existing populations.

I have mentioned that I believe the department of defense is limiting access to important technologies that could benefit the general public and contribute to survival system projects that would provide basic essentials in a difficult environment. Certainly fusion technology would be one such technology. At the same time, as far as I know the technology that is owned by the department is to complex and to expensive to become a mass marketed consumer product. The reason that I encourage people to get politically active along these lines is that I would like to release more advanced energy technology that could be mass marketed. Oddly enough, the most advanced technology appears to be very simple on the surface, mainly because the actual driving force behind the technology is completely invisible. A technology that does not involve moving parts is also low maintenance. The consumer fusion products that I envision would provide an energy output that is limited to the needs of an individual or family. There would be no way to scale up such a unit once it produced, nor could the unit be turned into any kind of explosive device, the unit would simply provide a steady energy output given water or hydrogen as an input. Even though releasing this kind of technology would not really be a national security issue, that doesn't mean someone will not try to appose it and try to invent national security issues. That is politics; it isn't always pretty or true.

Special interest groups need to be understood on a person to person basis, and on a group by group basis. That is the only way to change things for the better peacefully. Certainly the oil industry is a special interest group. I am sure that most of the people in the oil industry are good people. I worked in the oil industry for a few years. I also recognize that a trillion dollar industry represents a lot of money and power. I am a realist and a conservative strategist. I will not assume that everyone in the oil industry plays by the rules of common law. I would not rush out with a technology and products that threaten an entire industry without taking some precautions first.

The best precaution to prevent knowledge from being put on a shelf is to give that knowledge away to a large number of people, publishing a book for example. Even though I do not present detailed designs in this particular book, since the book is intended for the general public, I do present the general case which can be used by people who know technology and have the resources to do their own research and development. The answers in this book can be used to develop any number of solutions to the problem of energy given a bright engineer and some development resources. Obviously, if a stable oscillating sphere of energy creates energy as it oscillates all that is needed is to tap that energy. The rejuvenation products on the web site do exactly that, raising the pH of water and adding energy to water. This is a relatively small amount of energy, though an ideal amount for bio systems. These products represent a proof of concept that can be used to develop more advanced systems.

A crystal product has been developed and tested that releases enough oxygen from water to leave a form of liquid hydrogen at room temperature that burns as well as gasoline. Hydrogen can be held in liquid form at room temperature when spheres of energy larger than molecules hold the hydrogen together. Gasoline is just contaminated liquid hydrogen. The carbon in gasoline does not provide energy; it just goes out the tail pipe as exhaust. It is the hydrogen in gasoline combining with oxygen that provides the energy. That is why an engine running on pure hydrogen produces a little more power than a gasoline engine.

Trying to release that technology when no one but myself and a partner or two know about it would make it very easy for a trillion dollar industry to bury everyone plus the technology. That is not a statement of paranoia. There are many ways to destroy someone or something, the "best" methods appear completely accidental, natural, or just as an unfortunate sequence of events. The best way by far to suppress energy technology is to convince everyone that it is impossible to create energy and then call anyone who would talk about such things crazy, reducing the chances of financial support for that kind of project. Fortunately, I do not need financial support beyond what I already have combined with the help of a few trusted business partners. The political issue remains. Make no mistake about it, restructuring energy and changing the global power structure is a political issue. If the general public thinks that they are suddenly going to become energy self sufficient, without tens of millions of people marching on Washington to make it so, people are far to naïve.

It is no accident that everyone is economically tied to centralized systems delivering energy. This is true for individuals and companies.

David and Goliath is a good story. Given something as important as energy it is best to improve the odds going into the battle. If the general public clearly understands the workability of new energy technology first, that would make it much more difficult for a few powerful individuals to bury such technology before it can be shipped to people. Also, if millions of people begin to realize that it is possible to create energy, then even if my team was prevented from bringing something to market, maybe someone else would succeed.

There are a couple different design options for energy systems. The hardware needed for a completely self-contained energy system would be more expensive than many people could afford, at least for some years. By completely self contained I mean that the system would just run for decades without needing anything, besides an initial charge of hydrogen fuel if the system involved fusion of hydrogen up to helium. The mass marketable solution I envision involves a crystal catalyst added to water. The unit itself would be in the ballpark for consumer pricing, similar to the cost of an internal combustion engine for a car. The energy cost would be much less than gasoline, comparing the cost of the catalyst required to convert 1 gallon of water versus the cost of 1 gallon of gasoline. The engine or fuel cell would be completely clean burning, emitting only water vapor.

What would happen if we started using a different energy technology? How about all the people that are employed by the oil industry? We would need a rapidly expanding economy to employ all those people in other industries.

If the existing energy industry does indeed influence politics and government agencies, which I believe is the case, it would be an up hill battle introducing something new until the political landscape is changed. So, I prefer to focus on changing the political landscape first, rather than engaging in a battle that could be lost before it begins. If people really want a change they will knock on their neighbors door and say that something new has arrived in the world and that things do not have to continue to be as they have been. Until people start down that path I can only assume that the nation is not really serious about moving away from oil and is still "under the influence" so to speak of oil.

The oil industry is anything but stupid. Believing otherwise would be a mistake. The leaders of the oil industry know that when there isn't any oil left to sell, they will not be able to make money selling oil. That is my one hope for the oil industry embracing change. The industry may move on to new things after realizing that change is inevitable. Lets take a close look at the global oil situation.

Here are the statistics. So far we have burned 900 billion barrels of the easiest to find, easiest to pump oil. New oil discoveries per year account for less than 10 percent of the oil used per year. That new discovery number is dropping rapidly to zero. The technology used to find oil is extremely advanced at this point. The oil doesn't have much chance of escaping detection.

The total estimated remaining global reserve is about 1200 billion barrels and that includes found and unfound reserves. Why do countries count unfound reserves? When a country estimates how much oil is in the ground that number is used to secure international loans. Would a country lie on an international loan application? The reserves are simply "unfound", they are not actually lying, the banks give the countries some leeway.

A more realistic number for remaining global reserves is about 900 billion barrels. Those 900 billion barrels are harder to pull out of the ground than the first 900 billion barrels was. We continue to use more and more oil every year and that trend has to continue if want to grow the economy before switching everything over to a new fuel source. Switching everything to a new fuel source will take years. The trouble is that oil wells are now being sealed off at an increasing rate after achieving an energy break-even point. We are passing the peak of oil production on this planet. That means oil will only continue to get more expensive from here on out unless significant action is taken. Even if there is 900 billion barrels of remaining reserves, that does not mean that we can pump 900 billion barrels to an end user.

We burn oil to produce energy to pump oil, and we burn more oil to deliver oil to an end user. As oil becomes more difficult to recover there comes a point when the energy used to recover and deliver the oil equals the energy content of the oil itself. At that point the oil well is closed off. That does not mean that whatever oil actually remaining will cost more. It means the oil will never be pumped and delivered. No matter what the price of oil, if a company has to buy 10 barrels of oil to pump and deliver 10 barrels of oil the oil is going to be left in the ground. What if we used alternate energy sources to pump the oil? That would mean that the alternate energy source costs less than the oil, so we are still back to leaving the oil in the ground.

The price of oil is carefully controlled to prevent commonly known alternate energy sources from becoming competitive. More importantly, really competitive technologies like nuclear fission has been damned in the press and made to appear very dangerous and undesirable. Who do you think Green Peace is working for when they protest fission power? They are working for oil companies, and they are not helping very much when they do not present their own workable energy solutions. Solar power sounds very good. It is not a workable idea if the sun is blocked by ash for a number of years, nor is it a workable idea if half the land mass of the U.S. would need to be put under solar cells.

The entire energy infrastructure of the United States is built on a nonrenewable resource that is running out. It does not take running out to crash the economy. A 300 percent increase in prices would crash the economy, having no other energy infrastructure to fall back on. Natural gas is not going to prevent economic collapse if the existing energy infrastructure consists of internal combustion engines that run on gasoline, not natural gas. Fortunately, I am quite sure there is a renewable solution that will work with little change to the existing infrastructure. Hydrogen gas would be renewable, but it would also require significant new infrastructure.

Clearly, running low on oil before a completely new infrastructure is in place would be a problem. It would take many years to build a whole new infrastructure. So, what are people waiting for, and when will oil run out? That is easy to figure out. The amount of oil used globally per year is known, and global reserves are also known with a margin or error.

Let's assume that global reserves are 900 billion barrels, accounting for a 30 percent exaggeration by nations. Assume that 400 billion of those barrels can be profitably pumped and delivered to an end user. This estimate is conservative and should result in a safe estimate of the number of years remaining before an energy replacement is desperately needed. The U.S. consumes 25 percent of the world's oil at a rate of 20 million barrels per day. This is 7.2 billion barrels per year used by the U.S. alone. That is 25 percent of global usage, so across the planet we use roughly 29 billion barrels per year. Most of the profitable oil in the U.S. has been pumped and consumed, so the U.S. relies on oil imports. The entire Alaskan reserve we are talking about opening is only estimated to contain between 3 billion and 12 billion barrels. The flow through the Alaskan pipeline peaked less than ten years after beginning operations. Opening the Alaskan reserve would increase the flow through the pipeline for a couple years. It really isn't significant compared to the oil left in Iraq and Saudi Arabia. That is where most of the remaining oil is.

Let's calculate with 29 billion barrels per year, assuming we get very efficient and global economic growth doesn't raise the demand for oil. Four hundred billion barrels divided by 29 billion barrels used per year leaves us with a little more than 13 years worth of oil. A large percentage of that oil is going to personal transportation, commercial shipping, and agricultural equipment using gasoline. These products are all internal combustion engines of some kind or another.

It would be no small task replacing all these internal combustion engines with something different. Centralized electrical power would not be such a difficult thing to move away from oil. That has already been done to a large degree with coal burning power plants and some nuclear power plants.

The U.S. really do need to start converting everything to a new power source. Hydrogen is the best alternative because it is clean and it will work in existing internal combustion engines. In fact, there is a little more energy in hydrogen than gasoline so an engine converted to hydrogen would get a small power boost. The ideal solution would be a new form of liquid hydrogen, made from water, which would run in existing vehicles without having to rebuild the entire engine.

President Bush proposed spending 1.2 billion dollars on hydrogen energy research. He received a big round of applause. Putting some money toward hydrogen research now is better than doing nothing. It isn't much better than doing nothing. That approach demonstrates either technical ignorance or a hidden agenda. If you do not want to get anything done other than people sitting around in a lab researching, then pay people to sit around researching. If you want a product, fund a venture that is ready to build and deliver a product.

Here is a description of how matching venture funding could work. We identify a team with a good plan and a working prototype demonstrating a hydrogen economy technology. Hundreds of under funded but promising companies, groups, and individuals do exist that are capable of delivering products needed for a hydrogen economy. A quick surf on the net demonstrates the truth in that. Investors also exist who would like to invest if they can make money. The investors can take care of due diligence. If an investor is willing to put 5 million into a hydrogen economy venture, the public should match that with public money to encourage a hydrogen economy. That would be better than wasting money on research that isn't really needed. It would be wise to make sure the money is properly spent by making sure new people are hired to participate in these new ventures. It is possible to waste 1 billion dollars on research projects staffed by a few people who talk a good talk but are really quite ignorant when it gets down to shipping a product. So far this nation has spent roughly 100 billion dollars of public money on kinetic fusion research that cannot possibly succeed. Those 100 billion dollars did not create many jobs.

Certain rules could be established to get the biggest bang for the buck doing matching venture funding using public money. Maybe we just put in 1 million of public money to start into many new ventures, and then do a technical audit of the different companies to see which ones deserve more investment. In general, the goal would be to make the best use of public money to jumpstart the most promising hydrogen economy ventures.

Researching hydrogen to death for the unforeseeable future is not a good solution. The important thing to realize is that people need products not research. Workable hydrogen technology is already available, mass producing that technology will make the end product cost competitive. The technology of fuel cells, hydrogen storage, and other related technologies will only get better working on real ventures that use mass production techniques. The U.S. can build a hydrogen infrastructure now. The U.S. can build hydrogen vehicles now. Before I go off on a hydrogen economy tirade, I want to make one thing clear. The nation needs to change to a hydrogen economy now. I would say that three times in bold print if it would help.

Hydrogen is a very flexible fuel. Hydrogen works in a combustion engine. Engines designed to burn gasoline may need a different tune up or partial rebuild to run on hydrogen but that is better than having to replace every engine. Hydrogen can also be used in fuel cells to produce electrical energy at a very high efficiency. Either way, hydrogen is a clean renewable fuel source. Not many people have an oil well in their back yard. Granted, a few people do, but even those people do not have a refinery to go along with it. My goal is to see that people are energy independent. Using wind power, people could produce hydrogen right now using electrolysis in their own back yard. Since we do not have a hydrogen infrastructure in place, producing hydrogen using wind power does not make sense. If fuel cells were being mass-produced, and vehicles were set to run on hydrogen, people would have a chance at energy independence.

There are a number of existing methods for hydrogen production. The immediate solution, based on available technology, is to build more nuclear fission power plants. The energy of nuclear fission can be used to produce hydrogen from water or other sources like methane. I honestly believe that the whole objection to nuclear power was created by the oil industry secretly funding groups like Green Peace and discrediting nuclear power in the press. I wouldn't be surprised if nuclear plants have been sabotaged in the past. It is not a problem safely storing a bunch of spent fuel rods. Technology exists to do that safely. If people really object, in the future payloads of spent fuel rods into the sun. The most important thing at the moment is preventing economic collapse and securing the future of mankind, even if Green Peace goes on a crusade to prevent pollution of the sun.

As this book goes to print, congress has approved federal loans for new fission power plants, one of which is slated to produce hydrogen. Nuclear technology has advanced since the last nuclear power plant was built in the U.S. more than a decade ago. Breeder reactors designs are available that are much more efficient. Using these new reactors the U.S. has enough uranium to last a billion years. This is a step in the right direction. This step still relies on centralized power rather than creating systems that would make individuals and families energy independent. It would be best to have both new fission power plants and local systems for individuals and families.

Fission technology and all the building blocks we need for a hydrogen economy are available now. Fusion technology also exists, stashed away in the Department Of Defense. That will take a presidential decision to release, even in a limited form. I do not see high-energy fusion technology going beyond the Department of Defense anytime in the near future. That is far more energy than any small group would need, and putting that in the hands of the public in the U.S. would mean that the technology also goes into the hands of anyone on the planet with the money to buy it. I would support the release of a lower energy product that could not be reverse engineered, or converted into a weapon system that would threaten the conventional forces of the United States.

I have heard that the U.S. government declares over 2,000 patents national secrets per year. Of course, if the patents were declared national secrets there would be no way to tell exactly how many were declared national secrets. If an inventor tries to patent something that is related to technology declared to be a national secret, the inventor cannot tell anyone about the design, much less build it or market it. The inventor loses the design and is subject to a long jail term if the inventor discloses the details of his or her own work to anyone, much less in a publication. That is just the way the game is played regarding technology of a sensitive nature. It would not be unreasonable to assume that such an inventor would be placed under careful observation to be sure that they are obeying the intent of the law regarding the protection of national secrets.

Personally, I believe that national security administrators have gone over board putting some things in the national secrecy category that do not need to be classified as such. Granted, certain things need to be protected. At this point the public is so far removed from what is actually going on with technology that the public could be compared to a cave man going about the cave man life style while someone in the tribe sits in the back of the cave with a laptop and ray gun. If anyone in the tribe somehow manages to stumble upon the ray gun, and insist upon disclosure to the other cave men, that curious sort could find himself or herself being eulogized after a "lightning" strike. All of this gets down to the fact that some form of local, reliable, long term energy source is going to be needed for individuals to survive in the future. More importantly, keeping the public in the dark regarding technology, the public is not able to make informed decisions electing leaders. It is possible to provide a lower energy power source to the public without really compromising national security. That should be done, and people should at least have a vague idea of what the U.S. is capable of.

It wouldn't be an exaggeration to say that any energy technology that could seriously compete with the oil industry, besides nuclear fission, could be classified as a national secret. Certainly anything along the lines of artificial gravity and things related to super conductivity are. Super conductivity is about not losing existing energy, which is much less important that the technology of creating energy or releasing it via a nuclear reaction.

If someone were to try and build something anyway, without patenting it, a wide array of laws exist that could be brought to bear. For example, the highest classification of illegal weapons one can possess are energy-related weapons. You would not have to hold your breath long waiting to see how long it would take someone to disappear if they tried to publicly market such a product today. As technology advances, it becomes easier and easier to make a death appear to be an accident or natural causes. I will not go into details, beyond saying that the things

I mention regarding energy fields pervading the body are known in certain circles along with how to create weapon systems that directly affect these fields.

Secrecy laws were written with national security in mind during wartime. With the cold war, and the war on terrorism, things have not changed. Certainly designs for things like nuclear bombs, that can only be used for mass destruction, should be kept secret.

One big concern at the national security level is that advanced technology will get into the wrong hands and be used against the U.S. That is a valid concern. If a product is sold to the public that product is subject to reverse engineering efforts by anyone who would like to steal the technology in other countries. Not every other country abides by international patent law.

It is possible to create a fusion generator that cannot be reverse engineered. That may sound unlikely, but it is true. Advanced technology can include components that are very difficult to manufacture. A physical component provides clues that can be used to recreate the original manufacturing process. A stable sphere of pure energy is very difficult if not impossible to reverse engineer. That is because the real thing of value is the device at the factory that produces the sphere of energy. Once containment is lost, that sphere of energy will just drop straight through the planet into the core (if the energy is not embedded in a solid). It doesn't have to be a big sphere of energy to run a vehicle or provide local power. A hydrogen bomb creates a huge invisible sphere of energy that drops down to the core. It is the shock of that creation that shakes up the magnetosphere more so than the fire ball itself. The size of that hydrogen bomb energy sphere would be on the order of billions of times larger than what might be used in a local fusion power source.

The executive branch administers secrecy laws and national security, so you would think the president would have some say in these things. Not necessarily. It is true that the president has the authority to change these things. However, an elected president is kept very busy during his or her limited term. It requires a very strong technical background to even get involved in decisions related to national security technology. The president probably never sees all those national secrecy patents. It would make sense to only disclose national secrets on a need to know basis. The president would never be directly involved in the actual development of a product related to national secrecy. Kennedy had the vision to bring our nation to the moon many decades ago. Kennedy was also shot. One might wonder if anything else was going on in the background related to Kennedy and that assassination. I believe that we need a nation of free and informed people. That would mean a limitless supply of competent leaders that can be elected to replace anyone that gets shot trying to do right by the people of our nation. As I mentioned, some things still need to remain nation secrets, I just believe that the line has been drawn at the wrong place between secrecy and disclosure.

The United States of America has many strong points, and many people of principle; most doing what they believe is best for the future of this country. Those actions will be flawed to the degree that each of those individuals lacks a complete understanding. Times change. What may have been in the nations best interests decades ago may not be today. The federal government consists of transitory elected officials, plus people holding permanent positions. The people in permanent position hold the real power, not so much the people who were elected. The constitutional amendment that limited the terms of elected officials took an enormous amount of power out of the hands of voters, putting that power some place else.

Putting better elected officials in office is only one part of the solution looking at creating a better civilization. It could very well be that large and powerful institutions exists within or related to government that have very little to do with the legally elected leadership. It would require taking a very close look at all departments of government to determine how true or untrue this notion is. Doing spring-cleaning, one should look under the couch and behind the refrigerator for dirt. It would also require legislative change. All elected officials should serve as long as people want them to serve, and as long as those people in office are willing to continue to serve. That is the way the constitution was written. Our founding fathers knew what they were doing. Term limitations were not part of the original constitution, and should be abolished.

All this talk about politics is directly related to the subject of energy. Unfortunately, it really is politics that is preventing public access to advanced energy technology, not the lack of technology. Until the public understands that and demands change, talking about advanced technology will be just that, talk.

I would like to see the speed and efficiency of the U.S. converting to hydrogen compared to the power of the 101st Airborne Division and their ability to deploy world wide in 36 hours. Until the whole situation with fusion can be worked out on a national level we should start building more fission power plants.

Reliance on foreign oil, and air pollution, are only secondary reasons to switch to hydrogen. The main reason is self-sufficiency. I would argue that the main reason advanced energy technologies are not available is to create an environment of social economic control. If people had a virtually unlimited source of energy at home what incentive would they have to continue to work within a broader economy? That is a serious philosophical question. I would guess incentive in that situation could only come from an honest spirit of patriotism, plus a desire to contribute to whatever projects groups are doing within the society.

A concern for the future existence of mankind should over ride any question about what people may or may not decide to do with their own free time. I see the events of the future being so challenging that it will be necessary to have a dispersed population, each capable of surviving independently. It just isn't going to be possible to predict exactly where magma is going to erupt, where tornadoes will hit, and where solar flares will be concentrated. We can say what parts of the country are likely to remain above the ocean. Beyond that things will be in the hands of God. Local renewable energy is the key to self-sufficiency. Energy can be used to power grow lights, pump water, separate oxygen from water, and any number of things that are needed in an extreme environment. Let's take a closer look at hydrogen.

Hydrogen is an attractive energy source because it can be made from water via electrolysis. It is true that right now it is expensive to make hydrogen via electrolysis, compared to burning an equivalent value of oil. Even so, 1 pound of hydrogen contains more energy than 1 pound of gasoline. Hydrogen is better than gasoline, and much better than batteries. It would take $20,000 worth of batteries to store the electrical energy in 1 gallon of gas, or a similar weight of hydrogen. Hydrogen can be stored safely in a metal hydride, which is a solid, or as a liquid given the right conditions.

Alternate energy sources, particularly wind, are very practical in many parts of the country. It makes sense to store the energy produced by a wind generator as hydrogen rather than buying batteries and carting around all those batteries. We do not convert alternate energy to hydrogen now because existing appliances and engines do not run on hydrogen. Fuel cells exist that convert hydrogen directly into electricity that can run existing appliances, but today fuel cells are to expensive. If fuel cells were mass-produced the cost of fuel cells would be reasonable.

If local power systems, vehicles, and appliances ran on hydrogen then a single portable fusion generator could be put to very good use. That generator could just plug into an electrolysis machine and crank out hydrogen from water to run all those appliances and vehicles. We can separate the appliance or vehicle from the energy source. A great deal of effort goes into building new appliances. It would be a mistake to build all those new appliances to run on goal or natural gas for example, rather than hydrogen. We are not going to find a renewable source of fossil fuels, so it is best to target renewable energy sources during the construction of new appliances and vehicles. Hydrogen comes from water. This planet has plenty of water. Hydrogen itself is the most common element in the universe. So, hydrogen is a good long-term energy solution.

We already know how to build hydrogen storage containers and fuel cells. We just need to start building them as best we can in quantity. Breakthroughs and process improvements will be made on real production lines much faster than they would be made in a theoretical laboratory.

In summary, public matching funds for ventures targeting a new hydrogen economy represents the best way to jump-start a hydrogen economy. Those public funds going into profitable ventures will create jobs and improve the economy immediately, provided some rules are established insisting on the creation of an appropriate number of new jobs for the venture funded.

That is how I view the near term future of energy. Energy is directly related to economic growth, or the lack of it. Energy drives every facet of economic production to a greater or lesser degree. Even looking at the service industry, those people need a place to live, transportation, food, etc. All those things require energy.

If we take the fast track towards long term, low cost, renewable energy, then rapid economic growth will follow in the wake of progress. If we do nothing significant, continuing to rely on oil, there is a growth cap at best, combined with the risk of economic collapse. These things are important. Immediate effective action is necessary. It is not unreasonable to demand a 30 percent transition to hydrogen within a few years. Anything slower and we risk doing far too little far too late.

An economic boom is the direct result of getting busy and producing valuable new products. Reducing the cost of energy, and transitioning to a clean renewable fuel source, will produce economic expansion. The entire economic engine will run more efficiently using better, lower cost energy. People will be proud to be working in a clean environment.

What the U.S. does now with energy impacts the survivability of all mankind. If the U.S. does not take the technological lead, is there any other nation on this planet that is capable of doing so? I do not see one.

Survival systems will require local renewable energy sources. Centralized nonrenewable energy leaves entire populations at risk.

10

Survival Systems

The first thing to realize about survival in an extreme environment is that advanced technology is necessary. Grizzly Adams would not last very long on the moon without a space suit. In many ways the future environment will not be as extreme as the moon, in some respects it will be more extreme. In any case, survival will require advanced technology that is not currently in the hands of the general public. Even though this will be a transitional environment, the length of the transition will span years, so systems need to be built that are designed to provide basic needs for a very long time with minimal maintenance.

Some survivalists have ideas about driving their SUV into the mountains and living like Grizzly Adams. I will present the exact nature of rapid environmental change. In some areas, particular near the coastline, that SUV is not likely to make it past the local neighborhood. If the SUV did make it out of that coastline neighborhood, the chances are the SUV will run out of gas before making it out of harms way.

The body of man and woman is not designed with things like thick fur, claws, wings, gills, or other basic essentials in the absence of technology. Mankind cannot dive and live off the oceans, as do whales and dolphins. Mankind is destined for the stars. Mankind must use technology and intelligence, not brute strength, to survive. We are about to learn first hand about the power of God and creation itself. Global events may be a boot out of the nest and onto a path of galactic manifest destiny.

After a reversal of the magnetosphere the environment will be cleaner, and the soil richer, than it has been in almost 1 million years. At full strength Earth's magnetosphere will provide greater protection from harmful solar and cosmic rays. It is difficult to describe the importance of the magnetosphere as it relates to health and vitality. There is a reason why people lived so much longer during the early days described in the Bible. After experiencing the health benefits of Vital Water, and the rejuvenation effects of the different crystal products, one begins to understands how it is that energy can directly benefit life. Things are a little crazy with the magnetosphere being on the edge of transition into a new current mode.

The challenge is to transition existing civilization and emerging with a sense of purpose, hope, and fellowship. Mankind can live in harmony with nature using advanced technology, rather than degrading the environment with the technology we use today. After a transition a new full strength magnetic di-pole will exist and life will thrive again bloom around the world. Compasses will have to be reset, though I am sure that will be the last thing on peoples mind when the protective blanket of Earth's magnetosphere is restored. As it is, with the magnetosphere at less than ten percent in some areas, things are becoming untenable for life. There is more to the increasing cancer rate than diet and exercise.

Let's look at each environmental change that would result from a magnetic reversal. These changes would exist during a transition of the magnetosphere. Things would go back to normal, with the exception of the North Pole being on the other side of the planet, when this transition is complete.

Solar Radiation and Cosmic Rays

While in a circular current mode, the magnetosphere will not stop radiation from solar flares and cosmic rays. That is because the electromagnetic field will shift 90 degrees coming in line with solar emissions rather than acting as a large spherical shield. Essentially, the magnetosphere will look more like a big whirlpool just like the sun does at the middle of a solar cycle. There will be no South Pole, just an electrically negative North Pole around the planet. Rather than repelling things like solar flares and cosmic rays the magnetosphere will tend towards focusing these rays towards the equator. That doesn't mean higher latitudes will be safe, it just means things will be a little worse radiation wise at the equator. For planning purposes, radiation shielding solutions will be needed for all latitudes. Detailed calculations have not yet been done to determine exactly how much shielding would be required.

Earth's magnetosphere is transitioning into a circular current mode as we speak, looking at polar vortexes and ring currents in the Van Allen belts. A rapid transition has not begun. A large electromagnetic event like a class X solar flare will trigger the actual reversal. As time goes by the amount of energy needed to act as a trigger reduces. That is why the peak of a solar cycle is significant. I suppose large nuclear detonations could be a trigger, but in my opinion it will be a long time before something as small as nuclear detonation could trigger an actual reversal. Long before then many class X flares will have hit the planet. A solar flare is a ball of fire three or four times larger than the planet. Fortunately, Earth's magnetosphere in its protective mode is 10 times larger than the planet. It would still be a good idea to avoid detonating nuclear bombs. The trigger that I speak of is like that last little snow flake that causes an avalanche. The planet goes along for hundreds of thousands of years pulling energy from the sun into the magnetosphere, at some point that flow must reverse.

Peak solar flare activity lasts for about 4 years during the middle of a solar cycle. We are fortunate that the level of solar flare activity was lower than usual this solar cycle that peaked during 2001. Solar cycles tend to go in pairs of two. Usually a low-energy cycle is followed by a high-energy cycle. The peak of the next cycle will be during 2012.

Solar flares are dangerous when they are not blocked by the magnetosphere. During the moon mission this was a real concern. That is why the Apollo mission was done during solar minimum. An astronaut walking on the moon would be killed by a solar flare. The space suit does not offer enough protection. The moon is at the outer limits of Earth's magnetosphere. The Earth's atmosphere would provide some protection. At the same time, hitting Earth's atmosphere with large solar flares will affect oxygen levels. Most solar flares are large enough to engulf the planet and then some in flame. The magnetosphere is ten times larger than the planet, which is why the magnetosphere is able to ward off solar flares. The polar ice caps do not contain records of oxygen levels during the last reversal of Earth's magnetosphere 800,000 years ago. The polar ice caps formed after the last reversal.

There are a couple factors that need to be considered to determine the exact level of radiation impacting the planet's surface. Those factors include the degree of shielding provided by the atmosphere, and the strength of the incoming solar flare or cosmic ray. Although the sun being blocked by volcanic ash sounds bad, it is actually a good thing during a magnetic reversal. Nothing alive wants to be in the path of a solar flare unblocked by the magnetosphere. Even normal solar emissions would be deadly over time without the magnetosphere.

The solution is radiation shielding. The solutions that I present in this book will be things that can be economically mass-produced. I am sure there are better solutions given more time and resources. Workability is the most important thing. Sand bags would be effective and very inexpensive. The military and the oil industry use portable metal structures. These living quarters are built within the space constraints of a 40-foot shipping container making them easy to ship by sea or land. I will talk about military survival systems in more detail soon. For now, the point regarding shielding is that sand bags on top of a portable metal structure would provide more than enough radiation shielding. A wooden mobile home would not work well since the sand bags would probably cave in the roof.

Volcanic ash blocking the sun will be a blessing rather than a curse. Ash will replenish the minerals and nutrients in soils. Ash will help shield the surface from radiation. Ash will protect the atmosphere to some degree. There is still a risk of oxygen depletion caused by solar winds impacting the atmosphere over an extended period of time. Fortunately, there will be no shortage of water. Given a good power source water can be separated into oxygen to breath and hydrogen to run appliances.

The crystals that are available on the web site do release some oxygen from water. That is what the bubbles are that form while the water is being transformed. A stack of crystals has been created that releases oxygen and transforms water at a visible rate. This Crystal Stack uses platinum chips embedded in every other crystal stacked in a circular pyramid shape. The intent of these products is to produce Vital Water faster, not to extract oxygen. Although these crystals release some oxygen, one breath of air represents a lot of oxygen. Other solutions will be needed. The best solution is an air scrubber. An air scrubber uses energy to separate the carbon from carbon dioxide leaving pure oxygen for breathing. The carbon could be recombined with hydrogen in different ways to produce a synthetic carbohydrate for food. That is one example of a closed cycle using energy as a solution for booth air and food. That example is an over simplification of the technology that would be required. I present it as an ideal goal. There would be more to the whole synthetic food project, including how to make it taste decent. This chapter is intended to present general solutions rather than trying to describe actual products that do not yet exist.

That covers the first environmental threat caused by a magnetic reversal; increased radiation reaching the surface from solar winds and cosmic rays. The next threat to deal with is volcanoes and earthquakes.

Volcanoes and Earthquakes

Volcanoes and earthquakes will increase. These will be major events that will shape terrain over days or weeks. All existing fault lines will move miles not feet. All volcanoes with any recent geologic history of activity will become active. New volcanoes will be formed. The vast majority of volcanoes are found on the ocean floor. These under water eruptions, and movement along fault lines, will affect the ocean and coastlines.

The effects of earthquakes and volcanoes are being seen more and more often these days. The U.S. has been very fortunate over the past couple decades. The U.S. has not experienced the destructive force of a tsunami hitting a coastal city. Earth shaking, sink holes, avalanches, and landslides are common threats related to earthquakes. A number of other concerns come into play looking at a large number of major earthquakes. Major earthquakes can destroy things like dams, and disrupt land based energy and communication systems. Persistent major earthquakes would make repairing these systems that rely on long physical conduits. Telephone poles, under ground cables, and oil pipelines are vulnerable. All it takes is one break in a centralized line to cut off service to a large number of people. The Alaskan pipeline was designed to handle a magnitude 8.0 earthquake with gaps of about 20 feet opening in the earth below the pipeline. The pipeline barely survived a recent 7.5 magnitude quake in Alaska that opened a 15 foot trench under the pipeline. The Alaskan pipeline is able to stretch like a slinky. Most existing system are not as well protected as the Alaskan pipeline.

The important point is that centralized systems for food and energy distribution cannot be taken for granted. We are just going to have to change to local solutions.

Tsunamis are very large waves caused by earthquakes and volcanoes underwater. I expect a Tsunami event to raise the level of urgency in the U.S. Unfortunately, people tend ignore a crisis until it is upon them personally. That is what makes a pole shift so devastating. There is very little warning. A civilization has to advance to the highest levels of understanding to even conceive of this threat, much less prepare for it. How many people ignore their health until they are diagnosed with something bad like cancer? This nation must break this habit of doing nothing until things get so bad that there is no choice. Looking at a magnetic reversal, it will take about 3 weeks for the entire magnetosphere to shift and for all hell to break loose. On that day it will be far to late to do anything for anyone who is not already prepared. A week before that day things could very well be business as usual.

If it takes the destruction of a few cities by tsunamis to wake up the rest of the nation, those people will not have died in vain. I do hope that the people in California will respond quickly, get politically active, and support projects like getting personal transportation off the ground and into the air. That is the only way to protect a city from an 800 mile an hour tsunami providing only one or two hours of advanced warning as that 100 foot wall of water moves across the Pacific Ocean. There would be time to jump in an air car and hover up 200 feet.

Not more than a few years ago a tsunami hit several villages in the South Pacific killing 3500 people. If that same tsunami had hit Los Angeles the number would have been many millions not many thousands. A tsunami moves across the ocean at 800 miles per hour. We have no method of accurately predicting the occurrence of a tsunami. A major earthquake event in the Pacific Ocean could send a 100-foot high wall of water towards LA at 800 miles per hour. It is very difficult to imagine the destructive force such a fast moving wall of water would have on a city like LA. Everything in its path would be destroyed.

Volcanic eruptions produce large amounts of magma. There isn't much one can do about magma except get out of the way. Fortunately, volcanoes are much easier to predict than earthquakes. Earthquakes are caused by large flows of magma in or below the earth's crust. When that magma breaks through we have a volcano. Magma follows rules of fluid dynamics. The magma is always going to take the path of least resistance. That means that earthquakes will always precede volcanic eruptions, allowing major eruptions to be predicted by measuring earthquakes. Weak points in the earth's crust are known, which also helps to predict volcanoes in advance.

Over the past 200,000 years the planet has seen 4 ice ages. At least one researcher, the author of "Not By Fire But By Ice" predicts a new ice age triggered by the next pole shift. The idea is that ash covering the sun would result in lower surface temperatures, and all the water evaporated by under water volcanoes would fall as snow and ice at higher latitudes. That is certainly possible. I believe that the net outcome is going to be more temperate regions around the planet after a reversal. The last four ice ages were caused by magnetic excursion rather than complete reversals. It would be best to have heaters and mobility built into a shelter. Even if the net planetary result many years after a reversal is more temperate temperatures, one should expect weather extremes during a transition.

At this point, you may be getting the idea that not many people are going to survive future events on this planet. I wish I could say otherwise, but the truth is that even if the U.S. had many decades to prepare a significant percentage of existing populations would still be lost. That is why looking at things from a spiritual perspective is so important. The challenge is to maximize the chances of as many people as possible in geographically distributed local groups using advanced technology. By joining each of those groups in a new civilization that values life and recognizes the importance of teamwork, the United States will truly become one nation under God led by a great democracy that measures its strength by the power and values of each individual. It will not take very long to restore civilization to the population levels that exist today, considering that most people will not age as people do today.

Air Quality

The atmosphere itself may need filtration or oxygen enhancement in some areas. After this transition air quality will improve over time. This is a natural planetary cleansing process. It would be wise to prepare airtight shelters, or over pressurized systems with air processing equipment. Air scrubbing systems may not be necessary, but it never hurts to be better prepared than necessary.

We have covered radiation, earthquakes and volcanoes, and air. These are the key threats that a magnetic reversal presents. An intensification of existing weather phenomena can also be expected. Now we can consider survival systems to meet those threats.

Survival Systems

Clearly, the Grizzly Adams approach is not going to work until after this transitional event has passed. Advanced technology is required. Let's consider advanced technology related to basic essentials. The basic essentials are food, water, shelter, clothing, air, and mobility. There will be plenty of water, though it may be necessary to catch the water in tanks or to melt frozen water. I add mobility because some areas may turn out to be the wrong place to be in, even if the location was previously thought to be a good location. Saying what is the best place to be in advance will not always be possible. I suppose someone could argue that this planet is not the right place to be. There is truth to that, however developing space travel technology, much less building enough craft for a significant number of people, is not realistic in the near term. The survival system solutions that I will present are earth-based solutions.

I hope that the lack of physical preparation to date on the part of Christians is not because people are expecting to be whisked off by UFO's. The Bible does talk about God being with man creating a new civilization with man here on earth. I don't see anything in the Bible about alien races and UFO's. More likely the words "taken up" should be interpreted spiritually. Even if a physical interpretation is used, air rescue technology does exist.

Some people choose to interpret the Bible to suit their own purposes, or to cater to their own beliefs about how things "should" be. Evacuation by UFO's or by magical transport into the ether is certainly creative thinking, but it isn't what the Bible says. More likely this interpretation is forwarded by people who are lazy and do not think they should have to do anything along the lines of creating a new civilization, or saving themselves physically. Free will does exist. I believe God would be more concerned about spiritual existence than about seeing how many people can be stuffed into one aircraft or spacecraft. I believe that God inspired prophets to write about the challenges that mankind will face, hoping that some people will prepare much like Noah prepared for the flood. At the same time, I believe that the actual number of people who physically survive these events will be in the hands of individuals. Just as a good manager does not stand over the shoulder of every person he or she manages, I would not expect God to try and direct every detail of the lives of every individual. It is enough to tell mankind what the challenge is and expect that mankind will rise to that challenge.

If the majority of mankind were to fail in the eyes of God, and go down a path of complete annihilation, then I would not be surprised to see a few people inspired to prevent complete destruction. It is not enough for a minority to believe and want to take action. Democracy is the only fair and workable form of government. Even if the majority fails, the majority will at least learn from their mistakes, and that is what is important from a spiritual point of view. In a dictatorship the general population is never given responsibility, resulting in a population that only places blame as opposed to a democratic population gaining knowledge. As long as the general population continues to try and blame politicians, rather than electing better politicians, a true democracy will not exist. A population that blames others is a population leaning towards dictatorship.

Here are some survival system ideas to provide basic essentials like food, shelter, clothing, air, and mobility.

Shelter

Today people build large houses to hold all their stuff. The idea of a survival system is to focus on essential stuff and simplify. Today we rely on centralized distribution of energy and food. The idea of a survival system is self-sufficiency. The cost and effort that normally goes into building large houses with a bunch of stuff should be redirected towards smaller self sufficient living quarters with high tech energy systems, food systems, air systems, and shelter mobility if possible.

There is an industry that exists focused on converting standard 40-foot metal shipping containers into living quarters for the oil industry and the military. These living quarters are designed with utility and safety in mind, with options for doors, windows, AC, etc. The size is about the same as a mobile home, and the concept is the same, except the housing unit is constructed of metal and can be pressurized if necessary. Sometimes the military will deploy a mobile shelter as a command center in an environment subject to nuclear, biological, or chemical attack.

Our transportation industry is geared to move containers across the oceans and deliver containers by truck. These containers are a standard size and weight. Many of the semi trailers that you see on the road are just an ordinary 40-foot container on wheels. Since mobile homes are pulled behind trucks, the size restrictions are very similar. Consider these shelters a high-end military version of the mobile home that will survive a direct hit from the worst of hurricanes or tornadoes.

It is easy to install pre-designed windows and doors without compromising the airtight or watertight integrity of the shelter. Metal can be precisely welded and drilled unlike porous wood or concrete.

All sorts of accessories can be applied to a metal shelter simply by cutting holes and welding equipment or contact points in place. Extremely high winds turn ordinary objects into missiles. Those missles would barely dent a standard gauge steel shelter. Interior insulation could be applied to provide sound proofing. A container used as a living quarters would probably be empty enough to float. Dropped onto a typical flat bed truck, this kind of survival shelter could be moved to a safer location if necessary. If extreme flooding were a possibility in a particular area, one could attach some kind of additional floatation to the bottom corners of a shelter so that it would float right side up. Add to that some strong contact points, and the shelter could even be recovered using a winch or crane. Air lifting such a shelter would be possible, but that would require heavy lift aircraft.

A metal shelter would provide some radiation shielding. Additional shielding could easily be added using sand bags or any other typical shielding material including lead. The metal structure would provide sturdy enough support for these additions.

The military uses over pressurization in fighting vehicles like the M1 tank to provide protection in an NBC environment. To be on the safe side it would be best to have a shelter with air conditioning and pressurization. If the inside of a shelter is slightly over pressurized because of a concern about the air outside, an air lock of sorts is needed to safely exit and return to the shelter. A small interior "front porch" would work.

All of these design challenges have already been addressed by the oil industry and the military. Oil drilling operations result in the release of combustible gases like methane. Often these gases cannot be stored, burned, or completely contained on a drilling platform. Systems are built to collect the gas and burn it in the air over the rig as much as possible. Electrical systems can generate sparks. If you pull a plug out of the wall quickly you will notice a spark. That spark could be deadly on an oil platform. That is why the living quarters on an ocean based drilling platform are over pressurized. Clean air is pumped in at a pressure slightly greater than atmospheric pressure to purge important areas. If there are any leaks in the airtight area the clean air leaks out pushing away any air that might contain methane. Leaks are usually minor, and acceptable, since this results in a gradual flow of new clean air into the living quarters.

The military uses a similar over pressurization concept in the M1 tank and M3 Bradley Fighting Vehicles. Over pressurization keeps chemicals, biological agents, and radiated dust outside the vehicle. The same concept could be applied to a survival shelter. If a small room were built at the entrance one could enter that air lock, or porch, then purge the smaller volume of air in that area after closing the outer door. After purging the small porch, the entrance to the main living quarters could be opened safely.

Standard metal shelters geared for shipment on existing ships, trains, and trucks represent a practical economical solution for survival in an extreme environment. The cost would be higher than a mobile home, but probably lower than an average house.

Building a survival shelter is half the challenge. Selecting the right location for it, and securing the land, is also important. Higher elevation areas some distance from the coast, and away from fault lines and existing volcanoes, would be a good place to start looking. Many Midwest states look promising. It will be found that the federal government owns many of the best locations designated as natural parks, national forests, or bureau of land management (BLM) public land.

It is important that national parks and national forests remain in the public domain for future generations. If the land were sold to individuals it wouldn't be public land anymore. Offering that land to the private sector could result in a few people with many billions of dollars buying it all up. Some public lands can be leased for a period of time, rather than sold, so that the land itself remains a public asset for future generations. Right now land on national parks or national forests cannot be leased, only areas designated for mining or oil exploration can be.

A self-sufficient shelter using a clean renewable energy source should not adversely impact the environment. If that shelter is mobile it could always be moved to a different location in the future, returning the public land back to the way it was. Most importantly, if personal air transportation is used roads would not be necessary. Roads create a major impact on eco systems dividing up natural habitats. Granted, in the event of pole shift there would be very little left of existing eco systems. Even so, new legislation would have to be passed as preventative legislation in the light of existing eco systems and future eco systems. In other words, any argument put forward to change the way public land is used would be easier to pass through congress if it made sense no matter what the future holds. An argument can be made to lease public land if the person leasing is living in complete harmony with the environment, and if the structures moved in are not permanent.

The United States is fortunate having a significant percentage of mountain property in the public domain. If all that land had been privatized, the concentrations of wealth created by laws like the Federal Reserve Act would have resulted in all that land being owned by a handful of individuals. It is much safer to control large amounts of money than it is to wave around a flag bragging about having lots of money. The only benefit of money is the ability to spend money. There is big difference between technically owning money and being able to spend money. For example, if someone had 10 billion dollars in the bank, and another unnamed individual had a gun to that persons head directing the spending of those 10 billion dollars, who controls the 10 billion dollars? The names of individuals who control enormous amounts of wealth are not published in Forbes magazine. I am not saying the Bill Gates has nefarious connections. Bill Gates has done virtually nothing with his money beyond a few token projects relative to his total value on paper. Bill Gates does not control 30 billion dollars, that money is theoretical. If Bill Gates were to cash in much more than 10 percent of his equity he would crash the other 90 percent. I am saying that a few private individuals exist that control the spending of 100's of billions if not trillions of dollars, and those individuals are not worth that much on paper.

As it stands, the option remains to lease public land in the public interest if people decide that land is a good place for a shelter. A survival shelter would need a self-sufficient source of energy. I am sure centralized utilities will do their best to keep systems operational. Keeping power lines intact could be very difficult. Producing hydrogen at a centralized facility would make sense. The hydrogen could then be transported by ground or air. A hydrogen distribution system not relying on pipelines would avoid a single broken line leaving many thousands without energy. It would still be best to have energy systems included with a shelter, similar to the ideas just discussed in the last chapter covering energy.

Food

A significant long-term reduction in sunlight would cause food shortages. Since it is very difficult to predict exactly how long it would take to restore normal growing conditions, it would be best to plan on long-term solutions involving artificial light sources.

There are people making money multi level marketing algae as a food source. Some strains of algae represent a fine food source purely from a nutritional point of view. I don't know how it tastes. I have only tried a few capsules. People could get creative with spices and things, making algae burgers, algae bread, or whatever.

Algae grows quickly given water, minerals, and light. There will be no shortage of water, or minerals in the form of ash. It would be possible to build an appliance that uses artificial light to grow and harvest algae after throwing in some water and minerals. People's general health would really improve. This is only a temporary solution that would be used in a worse case scenario until the environment is back to normal.

Technology now exists to grow plants at a much faster rate. Using high pH water produced using a special crystal all types of plants will grow faster and healthier. In addition to better water, other natural technologies can be fielded to help with new agricultural methods involving artificial light sources.

Like Noah, the U.S. should have national repositories for various seeds and special shelters for some wild life particularly live stock. All these things take time to set up. Doing these things now shows that the public cares about mankind having a future. Not doing these things, the entire future of mankind and most wild life is left at risk. A pole shift is not the only risk. Personally, I know a pole shift is a near term certainty, not a "risk", but even those who doubt what I say cannot deny that mankind has other problems. Certainly people cannot deny that there is some risk of nuclear weapons being used as long as nuclear weapons exist in silos around the world. Certainly people cannot deny that a risk of terrorism exists, and that some body could come up with some nasty air borne virus or chemical capable of destroying most of the population. In that event, that only solution would be shelters like I have described until the threat can be analyzed and cleaned out of the environment. Chemical threats like nerve agent can be as bad as biological threats.

Mankind has come a long way over the past thousands of years. At the same time, we have created weapons that threaten annihilation of the race. Unless the whole human race suddenly decides not to go to war with each other, and unless every single individual with some influence on the planet becomes sane, history tells us that weapons are not kept on the shelf just because they happen to be very destructive weapons. It is time to put some efforts towards projects of creation and preservation of mankind, rather than doing nothing along that line while spending a majority of public money on more weapons.

That statement should not be interpreted as a dovish position. The U.S. will need a strong military in the future more so than it ever has in the past. It is possible that our country will face an extraterrestrial threat in the not to distant future along with environmental challenges. Maybe there is no extraterrestrial threat in this entire universe that would be even vaguely interested in this planet. At the same time, if there is, the best tactic would be to wait until the environment has done most of the work and then mop up and own a clean renewed planet with a bright future. It is always best to prepare for a worst-case scenario no matter how unlikely that scenario is. Doing so, it is impossible to be taken by surprise or become disappointed.

Up until 2003, and counting down from there, as a nation and as individuals very little or nothing has been done to provide for the future in the light of anything other than a best-case scenario. The sad thing is that even given all that we are blessed with today people still manage make a very difficult time of things. "Forgive them for they know not what they do." There is great wisdom in that, and maybe some hope if enough people take a stand behind setting things right. Before anyone decides to go on a crusade pointing out how wrong other people have been, realize that approach does not work. People have tried that approach throughout history, and guess where it go everyone? Right where we are today. The only thing that works is understanding the other person, forgiving them in their ignorance (in your own mind, not to their face), and then figuring out how to teach that other person at a level they can understand without talking down to them.

Back to solutions regarding food. When the body is working perfectly, in the peak of health, the body is very efficient. That is one reason why it is important that people understand their own body and how to achieve perfect health. Taking advantage of the rejuvenation products that are now available and creating a better body will be important for future survival.

The Star Trek food replicator systems would be a nice technology to have. Existing technology is a long way away from creating a synthetic plate of barbecue ribs that are as good as the real thing. Producing vital water really is a key technology related to synthetic foods. Having that solved, the next step would be creating synthetic carbohydrates, protien, and essential oils. That would take some serious development effort, but I am confident that technology is within reach. There is nothing better than the natural foods, breads, and meats that we enjoy today. A synthetic food could not hope to match that. Being healthy and satisfied, people could tolerate the change for awhile knowing that a better world and very long youthful life awaits in a future planetary environment that will be much better than it is today.

Clothing

So far we have looked at food and shelter. Clothing as challenging as other things, but it is still important. Today people spend a lot of money on designer clothes. How many people own a set of military quality protective over garments and a gas mask? I believe many people in Israel do. Not many people in the U.S. are than prepared. These things are not very expensive, and this one item could save the lives of millions. It would not hurt to have such things. Fires do happen on occasion, and it would be nice to know that one is prepared to escape a smoke filled environment for example.

Clothing is the least expensive survival system item. Protective clothing provides the biggest bang for the buck. Protective clothing allows an individual to survive in an extreme environment for a short period of time, possibly long enough to escape to a better environment. Building new shelters, appliances, and new transportation is expensive and time consuming. The price of particular items is important. Price reflects the amount of effort involved creating a particular product. The amount of effort a nation can expend over a certain time period is limited.

Mobility

Personal transportation needs to get off the ground and into the air. Even if there wasn't an environmental threat, or a need for quick evacuations, traffic is getting ridiculous. It isn't going to get any better on the ground. Traffic reduces quality of life and has an negative impact on GNP, considering all the hours setting in traffic. Inhaling low-level toxins in traffic is another problem, over many years all that bad air takes a toll. Carbon monoxide does kill quickly in a closed garage. I wonder how much different the environment is on a packed freeway during a muggy day in LA. I prefer to live outside big cities. There are solutions. It is just a matter of enough people standing up and insisting that those solutions get implemented. There will always be people trying to prevent progress by forwarding contrary arguments rather than solutions to better implement a desirable goal. People will say "what if some terrorist gets into an air car", or "it would be to dangerous with all those aircraft in the sky", etc. Life will never be completely safe. That is no reason to stop living, or settle for a low quality "safe" life living in fear.

There is a company called Moller International (MI) with a very cool prototype Sky Car. There is a link to Moller International on the book web site **www.answersforeverything.com**

MI's sky car works very well. MI believes the vehicles could be mass-produced for a retail cost of about $80,000 per sky car. A picture is worth a thousand words in this case. The sky car uses a breakthrough rotary engine design that reduces emissions by up to 90 percent compared to other internal combustion design. I am sure the design could be tweaked to run on hydrogen. Hydrogen contains ten percent more chemical energy than gasoline, so one would expect better performance using hydrogen. The sky car is a vertical take off and landing vehicle that can hover and achieve speeds close to 300 miles per hour. The two-seat version gets roughly 30 miles to the gallon.

The Sky Car is relatively safe. The vehicle has 4 engines. It can run on two engines as long as the two engines that are turned off are not on the same side of the vehicle. A built in parachute system is included that will lower the entire vehicle and passengers safely if necessary.

There are a couple things slowing down the release of the sky car. One thing is federal air traffic control system red tape. Actually, it isn't just red tape, the existing air traffic control system would need to be expanded. The second thing is that Moller International is under funded. No other company is even trying to do anything along the lines of mass-produced personal air transportation. That means that the public and investors really don't consider this project to be a priority or it would not be under funded. Even so, MI has struggle for over ten years, investing the equivalent of about 50 million dollars including engineering man-hours. The U.S. spent 1.2 billion dollars on the gearbox for the V2 Osprey. Investors and congress set priorities.

Building a national air traffic control system to handle millions of personal aircraft is possible. The system would have to be automated, more so than existing traffic lights. Aircraft would need to have transponders and computer systems to store air corridor information for a particular area. There is plenty of space in the air since the air is 3 dimensional. More air space exists at higher altitudes. Given high quality safety systems like parachutes, or auto rotation in the case of helicopters, things really would be safer in the air. One would not have to watch out for pedestrians.

Here is a rough sketch of an air traffic control system. Traffic would be strictly controlled in busy air space over cities. There would be main air corridors for travel at typical speeds over busy areas. To the left, right, and above the main air corridor, at the same altitude, we would have acceleration corridors used to enter the main corridor. The air space below the main corridor would be for landing and take off into an acceleration corridor. It would take some computer modeling over a city, combined with some practical experience to perfect the system. Building in an automated collision avoidance system would also be possible, since each aircraft would have a transponder and the ability to detect the location of other nearby aircraft. With 360 degrees of space to choose from to avoid an accident, I am sure that it would not be difficult to write software to automatically avoid accidents in the sky. Air space over less busy areas could be more flexible, as it is today with personal aircraft. The collision avoidance system would still be used. Existing personal aircraft would need to be upgraded to add a transponder and integrate with the new air traffic control system.

MI's air car is one good solution. There are many different ways to move personal transportation into the air. There are a few things that I like about helicopters.

Consider a hydrogen fuel cell powered helicopter with a super conducting electric motor / generator. This idea would also work with a traditional electric motor. Brushless electric motors are 95 percent efficient. The super conducting option allows the motor to be much smaller and lighter. Super conducting wire technology has come a long way. A Company called American Super Conductor was ready to go into volume production when I last called. The problem is that a market does not exist for mass production quantities of super conducting wire. The product can be manufactured in volume at reasonable prices.

A helicopter driven by an electric motor has some advantages over the sky car. Electric motors can always be turned into a generator by driving the motor using an external force. The large single blade of a helicopter could run through extreme environments containing clouds of ash, though there would be wear and tear on the blade. A jet engine on the other hand would tend to get clogged up. The only complete solution to that wear and tear problem is a gravitational propulsion system. That technology is not available now, and I do not see it becoming available in public quantities for at least a couple decades. So, an electric propulsion system is the next best thing eliminating the need for air filtration going into the carburetor, which could be a problem in an extreme environment.

The helicopter would not need the parachute system since helicopters can allow the drive shaft to spin freely and rotate down for an emergency landing.

The large blade of a helicopter attached to an electric motor could run backwards as a generator. The idea being that when the helicopter is not traveling it could be used as a wind turbine. During a magnetic reversal we can expect less sunshine and that rules out solar power. Wind will increase. With some creative mounting mechanism the electric helicopter could be turned into a wind turbine over night and replenish the onboard supply of hydrogen using electrolysis. The oxygen produced could be stored and used in a climate control systems.

Given a steady 30–40 mph wind that helicopter could crank out some serious power.

Using clean renewable energy sources, and air transportation, opens up many doors. It would be possible to live in harmony with nature using survival systems and transportation systems like the ones I have described. The system I have presented involve existing technology, things will only get better with advanced technology. I do not consider internal combustion engines very advanced, even compared to a primitive society using little more than fire and hunting with bows and arrows. On a galactic technology scale earth is still a primitive fire culture. Granted, the computer technology is a step in the right direction. If we live in harmony with nature, even improving the natural environment around us, why not open a percentage of public lands and national forests? These lands are a national resource that belongs to everyone. Leasing some land a reasonable rate would generate public revenue.

All the survival systems that I have mentioned are focused on individual and family sized units. The solutions I have presented are not prohibitively expense if the equipment and vehicles are mass-produced, nor would the effort required to build each vehicle be much different than the effort going into ground cars today.

That wraps up the technologies that would need to be built to provide for the basic needs of individuals and families in the face of a challenging environment. I am sure that there are many other solutions that would be worth looking at, including ocean based system and stockpiling solutions.

Ocean Based Systems

Ocean based ships, or semi-submersibles, are an interesting survival platform option. One advantage of an ocean based system is that it is isolated. One disadvantage is that large ocean platforms concentrate resources, rather than dispersing them. It is more difficult to shield a large ship than a ground based shelter. Submarines would be better, but the cost would be significantly higher.

It does not make sense to try and build ocean platforms for everyone, the cost or resources required would be prohibitive. All things considered, a land-based shelter away from the coastline would be safe and much cheaper. It would make sense to build some modern day Arks. These could be called Ark 1, Ark 2, etc. The U.S. nuclear submarine force is the nations most survivable deterrent. A submersible under 1000 feet of ocean water would be safe in any of the scenarios that I have mentioned.

As theoretically safe as a submersible would be, I admire the courage of individuals who would serve on a sub. I prefer open spaces, or at least knowing there is open space just beyond the walls, even if that space be outer space. It takes a certain amount of courage just to serve on a submarine.

A less expensive ocean based option would be a small floating city. There is actually a floating city project in progress. Some people in Europe got together to build a floating city to house about 70,000 people. This project is not survival system oriented, it is more of a luxury commercial venture. The project involves selling large multi million dollar suites. The designers do not plan to add any lifeboats. The argument against lifeboats is that the floating city will be unsinkable. Isn't it bad karma making claims like "unsinkable" after the titanic? Obviously, nothing is unsinkable, so why be unrealistic. The floating city does not include military defense. The navy builds some very survivable hulls, but we do not go around saying they are unsinkable. Why not just say the cost and space that would be taken by all those lifeboats is not acceptable to the owners of the floating city? Anyway, that is a brief description of a floating city project in progress.

A floating city survival platform would take into account the extreme environmental conditions that I have mentioned. A ship could be designed with these environmental considerations in mind. The focus would be on practical living quarters rather than large luxury cruise ship suites. There is nothing wrong with luxury suites in themselves; it is just that there is a time and place for everything. It would make sense building a modular floating city. The living quarters could serve as lifeboats if the quarters were designed with that in mind.

Stockpiling

I mentioned that stock piling is not a long-term solution. That doesn't mean that people should not stock pile food supplies. Just as the U.S. has a strategic oil reserve, so should we have a strategic food reserve. National agricultural output should be maximized while the sun shines to grow crops. What we do not use immediately, or export, should be placed in storage using some kind of food preservation system. Military Meals Ready to Eat (MRE's) have a shelf life measured in decades. Actually, MRE's are very good, I enjoyed them while on active duty.

Once a national stockpile of preserved foodstuff is created that stockpile could be used and replenished maintaining fresh reserve stocks. Shelf life would not be the main reason for drawing down and replenishing the reserve. A large investment in capital equipment and infrastructure would be needed to create this national reserve. Once the target reserve level is met we would not want to completely abandon that equipment and infrastructure created to produce foodstuff with a long-term shelf life. Drawing down and replenishing a certain percentage per year would keep those facilities maintained and operational. This project would also boost the economy. Actually, all the projects I mentioned would boost the economy. The economy is what people make it.

The economy is about putting people to work on good projects that improve quality of life. There is no excuse for political and financial leadership failing to create a robust economy, and improving quality of life across the boards.

Maximizing agricultural output would involve improving soil conditions around the country where soils are marginal for growing crops. Irrigation should be increased where necessary. The United States has an enormous wealth fresh under ground water. Natural aquifers exist under much of the Mid West containing far more fresh water than can be use over many decades. Even today, we have not made a significant dent in the aquifers, and the level is rising in some areas. There is a lot of land in this country that could yield some kind of crop, even if it is alfalfa, if put to use.

In the past farmers have been given subsidies to plow things under. I do not know if this is still true. If it is, we need to do an about face on whatever line of reasoning caused people to waste resources like that.

Survival System Summary

These survival system ideas are just that, ideas. Coming up with ideas is the easy part. Motivating 200 million to take action and implement grand plans is the hard part. Even if a majority can be motivated, getting 200 million people to agree on a specific course of action is even more difficult than motivating people to do something. That is what politics is about. Politics is not about wasted effort. Politics should be an important part of everyone's life taking an active role guiding the nation towards a better future.

11

Politics

Politics is a game of friends and adversaries. People with many friends have political clout. People without friends do not. Making new friends is a fun thing to do. The way to do that is really listen to and understand other people. Most people put up personal walls. There is nothing wrong with that; people are entitled to some privacy. The important thing to realize is that rejection should not be taken personally. Sometimes the person you would like to talk is just not available and that rejection comes more as an automatic response from a wall the person has put up rather being a well thought out personal rejection. Whatever the case may be, do not give up talking to everyone because one person has rejected you. There are limits to the number of close personal friends that any one individual can have. That limit is high or low depending on the person.

The types of relationships that drive national politics are similar to those that drive international politics. Even within a nation politics can degenerate into all out war. The U.S. civil war is one example. The civil war was an example of two large groups within the nation strongly disagreeing. Knowing that sometimes freedom has to be fought for, the federal government trains and equips armed forces to face any challenge. The military was evenly divided during the civil war putting the entire nation in danger on the broader international scene.

Consider what is at stake talking about national politics and electing leadership to represent personal interests. A nation is not unlike a big club. Everyone pays dues in the form of taxes, and these dues are either put in the bank or used to pay for club projects. It is the responsibility of the government to create money that is unique and can be used to represent goods and services for private and public exchange. That is the way the constitution was written. The federal government has the right to coin money. Things have drifted from that ideal after 1913 when the Federal Reserve was created. The Federal Reserve is a private corporation owned mostly by foreign nationals.

People should have a clear understanding of how money is created and added to the economy. The important net result is that enough money exists to represent the goods and services that people want to exchange. Just as important, new money needs to be created at a rate equal to the expansion of the economy. If people are willing and able to do new projects that increase GNP then money should be available to do the project.

The constitution gives the federal government the power to coin money. This is an important power. That power means that when GNP doubles, and new money needs to be created, that new money should be created and managed by the federal government for the benefit of all people. In 1913 the Federal Reserve Act was passed, greatly enriching 12 existing private institutions, and giving the federal government's power to control money over to those 12 private institutions. When it gets down to it, the 12 people running those institutions have far more real power and detailed control over the lives of every U.S. citizen than the U.S. government could ever have while the Federal Reserve Act remains in force. Money does represent real products and real labor produced by some individual. The amount of real money handed over to a private group called the Federal Reserve is equal to many times the existing U.S. government debt. People are welcome to vote, but the Federal Reserve will retain real economic power, and control this nation economically, as long as the Federal Reserve Act remains in effect. The Federal Reserve Act gives the power to distribute newly coined money to the Federal Reserve. Granted, congress votes to create some new money, but congress has little control over who gets the new money after passing it to the Federal Reserve. The Federal Reserve lends out the money to whomever it wants to, at whatever interest rate that the Federal Reserve sets. In return, the Federal Reserve gives congress a Federal Reserve note. The projects that the Federal Reserve wants to see funded get funded, not the projects that congress wants to see funded. Essentially, most of the economic stuff that represents the society we live in today is stuff that was approved by the Federal Reserve since 1913. Conversely, many of the things that do not exist, do not exist because the Federal Reserve did not approve loans for particular projects. Private banks under write all funding for companies doing an Initial Public Offering (IPO). Considering how many restrictions exist related to raising private capital, and the fact that private investors have to get the money to invest from someplace, one begins to understand how a very small group of people not elected by the public could determine the exact structure and future of society. Over the past few decades the path this nation has been going down has not been the right path considering what the near term future holds for this planet.

Here is how the Federal Reserve works. The Federal Reserve buys new currency from the federal government at 4 cents per bill. The Federal Reserve then owns the new money and can lend the money to whomever it wants to at whatever interest rate the Fed wants to set. The new money then goes into circulation. The Federal Reserve is required to hold a certain amount of government bonds related to the new currency purchased. Of course government bonds can be purchased with Federal Reserve notes making that arrangement circular. Rather than get into all the details of the system let us first focus on how the system affects everyday life through macroeconomics. If the Federal Reserve were to lend out all the new currency, and if everyone were to repay the Federal Reserve with interest, the net money in circulation would actually decline. Obviously that doesn't happen. The private individuals who own the Federal Reserve banks can do whatever they want with the new currency. In order to make this appear legal, and not as if the federal government is just giving away all new money to the Federal Reserve, the Federal Reserve is required to return most of the profits of the Federal Reserve corporation to the U.S. treasury. The money that goes back the treasury is spent through government spending adding net money into circulation.

Do you see the flaw in this system? Theoretically, the Federal Reserve could lend out 100 billion dollars to drug lords in South America knowing it will never be repaid and not be any worse off for it. The 100 billion dollars could simply be written off along with the 4 cents per bill paid to buy the new currency. A write off is not profit. Granted, on the books it would probably be best to describe the drug lords as some benevolent third world reconstruction project rather than say they are drug lords. All bad debts put more money into circulation reducing the net value of the other money held by the American people. This is called inflation. Section 7 of the Federal Reserve Act, paragraph A, provides for a 6 percent dividend to be paid to the owners of the Federal Reserve prior to paragraph B which pays money back to the U.S. treasury. That dividend puts more into circulation. More importantly, the Federal Reserve Act requires that the federal government use the money paid back to the treasury to pay interest on the national debt and pay down public debt. Federal bonds are purchased using Federal Reserve notes. The system is circular. Much of the money paid to the treasury comes right back the Federal Reserve.

The power of money is the spending of the money. It is wrong for the American people to give away that power to a few private individuals. For example, if GNP were to rise from 1 trillion dollars to 2 trillion dollars in a ten year period that means that the nation has produced 1 trillion dollars worth of new stuff that needs to be represented by new money. If no new money were introduced there would be a 50 percent deflation. That 1 trillion dollars worth of expansion represents the labor of all U.S. citizens. That new money should be coined and spent for the benefit of all people at the discretion of congress, not at the whim of a few private banks. In my opinion,Kennedy was assassinated because he tried to right this wrong and abolish the Federal Reserve System in favor of a real Federal Bank as described in the constitution. If Kennedy had been able to educate the majority of Americans on the subject of macro economics then Kennedy might have succeeded. If the people who assassinated Kennedy knew that 150 million people also knew the truth, and that those 150 million would just elect someone else as determined as Kennedy, then there is a chance they would have negotiate with Kennedy rather than simply shoot him and be done with the whole issue. Essentially, that is why I am writing this book, and why I am presenting the basics of advanced technology that most people would try to keep for themselves. If thousands of people know that advanced energy technology is possible, and if people have a sample of a lower energy crystal product that can be reverse engineered to help develop higher energy products, then I would hope that special interest groups realize they can't shoot everyone and that eventually someone will bring such products to market.

I cannot overstate how important it is for people to avoid ignorance and also make sure that neighbors also avoid ignorance. Do you really expect the people who established the Federal Reserve System, and who influence congress, to provide public education regarding the Federal Reserve System or macro economics? One begins to understand the sheer arrogance of the Federal Reserve reading the final paragraph of the Federal Reserve Act of 1913, section 31. As if stripping the federal government of the power to coin money was not enough, the final paragraph of the Federal Reserve Act says that the federal government has no right to repeal or abolish the Federal Reserve Act. That right is expressly reserve by the Federal Reserve. That is why the Federal Reserve is called the Federal Reserve. The Federal Reserve agreed to repeal Section 26, but beyond that the Federal Reserve Act of 1913 has not changed significantly.

An amendment to the constitution was never passed formally giving over all the power of government to the shareholders of the Federal Reserve, so the Federal Reserve Act of 1913 can be repealed including that last paragraph saying it cannot be repealed, which is the most unconstitutional and unpatriotic wording that could ever be conceived of for an act of congress that was not part of the original constitution. The problem is that the people who have worked their way up through the two major parties have agreed with the existing system, and after seeing what happened to Kennedy they are reluctant to do anything but agree.

The founding fathers were very confident that people were created with divine inalienable rights. These rights were written into the constitution along with the form of government. At the same time, the founding fathers realized that they were human and that a democracy might want to change things so the founding fathers gave congress the power to pass new legislation and repeal legislation as congress sees fit. The people who wrote the Federal Reserve Act of 1913 considered themselves Gods over the American people, and far more powerful than any mere founding father. Those individuals believed that they could create something that could never be change or repealed, essentially nullifying democracy. The Federal Reserve Act is an act of congress; it isn't even an amendment to the constitution. As long as the American people bow down to and fear the Federal Reserve God the American people will never be free of the economic chains they have allowed to be placed around their own neck. This issue is more significant than the tax on tea that started the American Revolution.

I cannot overstate the impact this has had on the quality of life of every single American. Every corporation that exists or does not exist, and the personality types chosen to supervise people in those corporations, where all carefully selected for funding by banks related to the Federal Reserve. Banks underwrite all IPOs. The level of financial stress and the rate of unemployment have been carefully created in this nation of bounty. There is no shortage of good projects that could be done. If the federal government had the power that our founding fathers intended then congress would be able to approve programs to fund thousands of entrepreneurs who have good ideas for doing things that would move this entire society forward and improve the quality of life for everyone. Entire books have been written trying to disclose the nature of the Federal Reserve. It would probably take an encyclopedia to list the most important crimes related to the Federal Reserve. The past cannot be changed, and I am not going to try and list specific abuses of power in the hands of the Federal Reserve. The one thing I will mention is that evidence does exist linking members of the Federal Reserve to both World Wars I and II. These wars were probably started by individuals secretly fanning the flames of intolerance on both sides of the Atlantic for financial gain related to military spending.

It is time to get busy and elect independent candidates who can be trusted, and who are not associated with either major party. Both major parties have failed this nation allowing the status quo to continue unchanged. The issues that the two major parties are now arguing about are little more than distractions keeping the attention of the American people off the things that really matter. Personally, I am "pro life". At the same time, I realize that the pro life issue will never be resolved while special interests groups fan the flame from both sides knowing that people will focus on this issue at the expense of everything else. People that are "pro choice" still recognize that abortion is a bad situation. A compromise can be arrived at that includes education, family planning, and even state adoption if necessary, and if all these advance planning ideas fail. The rejuvenation products that are now available will allow a mother to return to better than new condition even after the pains and stress of bearing a child. I would not appose "the day after" birth control pills to account for extreme circumstances like rape, but invasive destruction of life is a just not the kind of thing that any mother wants to live with for the rest of her life. If abortion were not an option people would be more likely to act more responsibly to begin with.

Am I exaggerating when I say the Federal Reserve considers itself to be God? Consider that the founding fathers listed 3 inalienable rights called the rights to life, liberty, and the pursuit of happiness. Inalienable means that these rights cannot be taken away because the founding fathers considered that these things were divine rights conveyed by God. The Federal Reserve Act of 1913 claims that the Federal Reserve has an inalienable right to take all of America's new money and do with it as it pleases as described by the Federal Reserve Act. The authors of the Federal Reserve Act wrote section 31 saying "the right to change, amend, or repeal this act is expressly reserved". That is why it is called the Federal Reserve. In other words, the authors of that act believed that congress had no right to change the Federal Reserve Act of 1913 without the permission of the Federal Reserve. This is clearly a claim to a new inalienable right granted only to one particular private corporation led by one particular private individual. If inalienable rights are divine rights given by God, and if the Federal Reserve can invent new inalienable rights, what does that make the Federal Reserve? Even the founding fathers left the inalienable rights listed in the constitution open to modification by constitutional amendment. Complete global domination in the hands of a few private individuals at the expense of everyone else is no "conspiracy", it is a black and white reality written into the very law of our land.

There is only one logical explanation for the existence of the Federal Reserve System. In the past some individuals with the power to make it happen decided that transitory elected officials could not be entrusted with the power to control money. The power to coin money is worth nothing without the power to spend the money. Why would an elected official give up the most important power of federal government to a private institution? There are only two possibilities. Either the private institution bought the elected officials, threatened the elected officials, or the elected officials did not feel capable of handling the responsibility that goes along with spending money. One scenario demonstrates a complete lack of integrity or courage, and the other scenario demonstrates a complete lack of leadership ability. What is done is done. Dwelling on the past will not change the future. The people in Washington need to be replaced by courageous patriotic individuals who are willing to put their life on the line for this nation.

The Federal Reserve Act of 1913 is a public document that can be viewed by anyone visiting **www.federalreserve.gov.** The document is buried in the web site. You will need to go to the site map and look under the Federal Reserve Board topic under which you will find a link to Federal Reserve Act.

I have heard rumors of "secret societies" or "the powers that be". How could a society that was secret have any impact on civilization, much less control everything? Control is about real influence. It is about funding companies like GM and not funding sky car companies. It is about funding oil companies and not funding anything else that might compete. Control is about denying loans for truly innovate new ideas, and supporting things like the existing housing industry instead. Control is about setting a particular unemployment target to manipulate the cost of labor. A secret society that does not affect day-to-day life might make for a good movie, but such a society would be impotent. The truth is that the majority of Americans have failed to understand economic theory, and that ignorance has been taken advantage of legally by a small group of people who do understand economic theory. That is life on a small scale or on a large scale. It takes two to tango. The few people who have tried to change the nation's monetary system did not take into account the fact that change in a democracy involves educating more than half the population. Change starts with education. After the majority of the public realizes that there is a better way of doing things a new bill has a chance of passing. Until then, any new bill will fail.

In all fairness, the leaders of the Federal Reserve probably consider that they are doing what is best in the interests of this nation. Remember, even the worst people usually believe they are doing right on some level. If the distribution of money had been left in the hands of congress the past 80 years it is possible that pork barrel politics would have ruled the day, and that the entire nation would have tended towards sloth rather than becoming the global economic super power that the U.S. is today. That economic infrastructure is important. What is needed is a retooling of that infrastructure to produce different products, and more efficient management of projects and labor. A centralized Federal Bank would accomplish that. Knowing the nature of future challenges, the popular majority can be trusted to implement a monetary policy that results in more economic expansion and better products rather that implementing something slanted towards sloth.

Many people wonder why evil and injustice are allowed to exist if there is indeed a God that cares. The first answer is that free will cannot exist without a real choice between good and evil. The second answer is more complex. How can a people of God learn to bring the joy of life and justice to a universe if those same people are blind to evil and injustice? Sometimes even evil has a purpose in the grand scheme of things. Defeating evil with compassion does help the unjust person learn that there is a better way to live life.

Any solution that is chosen to set federal monetary policy to rights must improve the economy immediately and not result in service disruption. A Federal Bank must be established prior to abolishing the Federal Reserve so that all business that the Federal Reserve would normally handle can be rerouted to the Federal Bank preventing any disruption of services. The Federal Bank should deal directly with companies and individuals. The goal is to get projects happening that are in the public interest. The only way to make sure that happens is to have federal loan officers. Each individual has a social security number, and a social security account, so much of the system is already in place.

On loan applications, payment of taxes and job performance should be the most important factors rather than credit ratings built up prior to the establishment of the Federal Bank. The Federal Bank would establish a new standardized credit rating system. Today there is so much confusion in the credit rating system with several different private companies and a myriad of different reporting requirements that the credit system is becoming a joke. A Federal Bank would be very efficient, using the Internet, mail, phone, wire transfers, and automated bill paying for all transactions. Many transactions could be automated. There would be no need to create a large infrastructure of bank buildings for a Federal Bank. ATMs exist for cash withdrawals. The vast majority of transactions do not require a face to face meeting, therefore things should be automated as much as possible freeing up peoples time to work on projects that are really important to the nation.

In my opinion, the existing credit and banking system works something like this from the viewpoint of the average consumer. The federal government gives a bunch of money away to the Federal Reserve. The Federal Reserve loans that money out at various interest rates. When loans are not paid, often because the Federal Reserve has mismanaged the economy, the loan is written off. The Federal Reserve doesn't care much about that since the money was virtually free to begin with and they pay 98 percent of all profits back to the treasury anyway. If the Federal Reserve were to disclose this attitude people might question the wisdom of the system, so the Federal Reserve hounds people to no end to pay bills that they often cannot pay because of employment conditions beyond their control. This makes life less fun to live than it should be for everyone, including the Federal Reserve, since if there is even a thread of decency in the people who run the Federal Reserve they must feel some remorse related to undisclosed manipulation of the American people. The honest consumer Federal Reserve write offs serve to mask much larger write offs from loans to individuals or corporations who are personal friends of the Federal Reserve. I am not claiming that any particular corruption exists in the Federal Reserve System; I am simply saying that the structure of the system does lend itself to corruption. If the Federal Reserve System sounds like insane monetary system realize that one great error people make is assuming that everything has to be sane, logical, or fair. Things are only sane, logical, and fair if they are made to be so. Since there can be only one truth, while at the same time there can be an infinity of falsehoods and non-optimum solutions, the odds do not favor truth and perfection.

If the American people are educated there is good reason to believe that an effort will be made to restructure the monetary system. Time can only tell. Unfortunately, politics requires money. Since the Federal Reserve influences the distribution of money, and the established major parties provide themselves perks like matching federal funds, the whole situation becomes a difficult situation not unlike a vicious circle. With matching funds a candidate can raise 40 million dollars and get a matching 40 million from the federal treasury. Significant matching funds are only available to the two major parties. A very small amount is available to minor parties based on the number of votes the minor party received in a prior election. I disagree with providing matching federal funds, but that is not going to change until the vicious circle is broken. If people become educated and determined to accomplish something then all the money in the world would not buy a single vote. Once again, the vicious circle comes into play since it takes money to educate and communicate with a large number of people. The only way to break the circle is a rapidly growing grass roots movement made up of people who realize what it takes to win. A grass roots movement of 20 million enthusiastic people is good, but if those 20 million people do not recognize that it takes money to reach another 200 million people in the nation quickly then that grass roots movement will fail. If 20 million people were behind a project and each contributed 20 dollars on the average that would be enough to outspend the total being used by both major parties.

The situation is further complicated by the unscrupulous use of money that can influence people who fear death or who fear for the safety of loved ones. The body will always be a transient thing, while an idea can last forever and shape the lives of many future generations for the better. It is a rare person who truly believes in death before dishonor. It will require many such leaders to bring about real change. Since this nation is a democracy it should be possible to achieve change peacefully, even if the death throws of an old system take down a few people dedicated to a peaceful transition to a better system. Historically, Kennedy was the only leader who might have been assassinated for trying to bring about political reform. At the same time, Kennedy might have been the only person who had the potential to create enough public support to change the monetary system. When the stakes are high people tend to start playing hardball. If nothing else the future is bound to be interesting.

I am sure that some existing private banks would protest the creation of a Federal bank. All bank employees depend upon the Federal Reserve System. Many of those individuals would need to find different jobs. A Federal Bank could employ some of those individuals if their attitude is exceptional compared to the average bank loan officer. A loan officer makes no contribution to GNP in the form of a real product or a service. A loan officer certainly doesn't provide any form of quality entertainment, philosophical enlightenment, or spiritual support. A computer can compare income, tax payments, and loan specifics then issue a payment by wire or mail a denial. It would be good to have a system for people to petition a human loan officer in the event that a loan request is denied. Loan officers should be the most humble people on the planet, yet they are often the most arrogant. The first will become last and the last will become first.

Abolishing the Federal Reserve Act, and creating a Federal Bank, is another one of those sensitive issues that could result in raising the ire of special interest groups that wield considerable power. The only way the American people will be able to change things for the better is by taking back control of monetary policy as it was written in the constitution. That will require grass roots politics and a no fear of death attitude in the face of long odds. My guess is that most of the people who currently hold office in Washington are sympathetic towards the Federal Reserve. Those individuals would need to be replaced if the lie of a private Federal Reserve is to be replaced by a real Federal Bank.

A Federal Bank could change the economy for the better by actively seeking projects that benefit the overall economy and create jobs. A new legal tender Federal Note could gradually replace Federal Reserve notes currently in circulation. A Federal Bank that services corporations and individuals would put the power to improve the economy back in the hands of the federal government where that power belongs.

For purely practical reasons related to the continuation of services without disruption the quickest solution would be to federalize the Federal Reserve System. One might ask how that would affect the federal debt. Federalizing the Federal Reserve System would only reduce the federal debt by about 10 percent, which is roughly the amount in government bonds held by the Federal Reserve as collateral against issuing Federal Reserve notes. The remaining 90 percent of bonds in existence are held outside the Federal Reserve System and would still be valid debts. Any citizen can buy government bonds, and these bonds would need to be honored.

There are two ways to pay the federal debt. The first way is to reduce borrowing from the private sector, and improve the economy resulting in more existing money flowing into the treasury that can be paid back out to reduce the debt. As GNP increases some new money can be created without causing inflation. This new money can be used either for new projects or to reduce the federal debt. This first way is the best way to pay the federal debt. The second possibility would involve the creation of new money ear marked to pay the bonds that represent the federal debt. This second option would involve significant inflation since the amount of money in circulation would increase without a corresponding increase in GNP. Inflation or deflation is a direct natural result of supply and demand. GNP represents supply and demand is represented by the amount of money in circulation that is being spent. So, inflation can also result from money being piled up in the private sector that is suddenly spent if GNP cannot expand fast enough to meet the new demand. In the grand scheme of things that kind of inflation is not particular bad since it does encourage a rapid expansion of GNP. The net result would be more products produced, and eventually the distribution of money would even out between labor producing products and whoever piled up the money to begin with. A more even distribution of money would end the inflation spike and make the monetary system more stable going into the future. These things are just likely results of natural economic forces rather than a specific monetary policy.

I began this discussion on money to answer how new money arrives in circulation. Over a ten years period given a doubling of GNP from 1 trillion up to 2 trillion dollars, without deflation, rough 1 trillion dollars of new money must arrive in circulation. Clearly, if 1 billion dollars was lent out then paid back with interest over that ten year period then deposited back in the U.S. treasury as profits are suppose to be deposited by the Federal Reserve then that would not result in a net increase of money in circulation. Even though the federal government should have the power to coin new money the power of distribution of that money has been given to the Federal Reserve. Therefore, a rational system to put new money in circulation cannot exist. Net new money arrives in circulation via a combination of consumer loan defaults, federal deficit spending, and defaults on international loans made by the Federal Reserve System. When the Federal Reserve makes an international loan that money is used to buy American products and services. In my opinion, of the three possible methods of bringing new money into circulation, defaults on international loans represent the greatest contributor. Determining that for a fact would require more investigative authority than I personally have. This is unjust not simply because Americans end up doing a great deal of unpaid work for the rest of the world, but mainly because the American people have no say in who will receive all that charity. Sometimes that charity does more harm than good.

Economics and politics are tightly linked. The basics of economics need to be understood first, and then an individual becomes able to judge the effectiveness of political economic management. Once a majority of people in a democracy understand the relationship between politics, economics, and quality of life people then become capable of electing leaders to change things for the better. No civilization exists in a vacuum. People must also understand the nature of both the challenges and the opportunities that exist given the existing level of technology owned by a civilization. A democracy really is a unique form of government looking at thousands of years of history. The average voter must take immense responsibility and be very intelligent for a democracy to result in a decent civilization. Not only that, each voter must be willing to educate every other voter who is complaining or ignorant. There simply is no room for complaint or mud slinging in a democracy. A democracy truly is government by the people. If the majority is acting foolishly that is just because a wise minority has failed to educate the majority. The minority will pay the price of that failure until the minority figures out how to work together with the majority and convince them that there is a better way to do things.

Knowledge is power, and ignorance lends itself to be taken advantage of in any way imaginable. That is one reason why I become concerned when I see entire communities in this nation, around Miami for example, of individuals who have the right to vote yet do not even speak the language of this nation. How can a voter make an informed decision if they cannot even read? Public programs do exist to help these people learn English, but while visiting places like Miami I find that some Spanish speaking people are making no effort to learn the English language and almost seem intent on establishing their own little foreign speaking state in the U.S. That is a problem for everyone including those people who are not learning English. All of the laws in the U.S., and most books dealing with education are written in English. It is difficult to move society forward without having at least language as a common denominator. English is a very rich language that embraces technology. English is the language of this nation. Just as the KKK is wrong, it is also wrong to establish any racist group, including Spanish racist groups. Teamwork and brotherhood with all people is not just nice thing to say, it is essential for the survival of a nation and maybe even for the survival of all mankind. Tolerating racism or unpatriotic attitudes hurts the racist and the rebel without a cause as much as it does the decent people trying to get along and change things for the better. Even racists and traitors are citizens, and an effort should be made to enlighten these people before they can pass a point of no return. If racist individuals become destructive then the only option a nation has left is the criminal justice system. Even that system is an attempt to reform in cases that do not warrant the death penalty. I would like to see white supremacy racist groups like the KKK start to reach out and help Spanish racist groups with a literacy campaign. That kind of thing would be the only acceptable amends for groups like the KKK to show that they have changed and are actually working to help others learn the language of this nation so that they may become informed citizens. The federal government should be promoting English language training rather than trying to develop a Spanish alternative for all services provided. In the long run it hurts both the Spanish individual and the society in general promoting anarchy of different languages within society. If what is being done today were carried to its logical conclusion there would be a dozen different choices to make for every phone call, and people would become more and more divided not being able to communicate directly with each other. English is a rich technological language. This book would lose something in any translation. As it stands today people who only speak Spanish in this country will not be able to read this book, and therefore will not gain a detail understanding of the information I present including how the Federal Reserve System works.

The Federal Reserve System has a certain beautiful complexity causing even most English speaking Americans to view the system with a sense of awe, lacking the ability to completely understand it. This sense of awe, and the fact the even people who try to describe the system as a conspiracy don't even understands it, results in a majority that cannot suggest a better way of doing things. For example, one conspiracy theory person with a web site argues that the Federal Reserve System is bad because it requires the federal government to pay interest on the very money authorized by the federal government but created by the Federal Reserve. It is not true that the federal government pays significant interest on bonds held by the Federal Reserve. Ninety eight percent of the interest paid on bonds held by the Federal Reserve is paid back to the federal treasury since it is counted as corporate profits. A system cannot be changed if people do not understand the existing system or the motivations behind creating that system. I have had to simplify the description of the Federal Reserve System in an effort to make the most important points clear. For example, it is not completely true that all of the power to spend money has been abdicated by the federal government. The profits that are returned by the Federal Reserve to the treasury can be spent and that does add some new money to the economy provided that the Federal Reserve reports a profit. It would be difficult to find an unbiased auditor with the ability to trace the truth of every transaction dealing with a private corporation as powerful as the Federal Reserve. Covering every nuance of the Federal Reserve System is not what I am interested in. The fact is that only the federal government should have the power to coin new money and put that new money into circulation. I am not interested in improving the Federal Reserve System; I am interested in abolishing it or federalizing it.

Logically, the federal government created the Federal Reserve System, and the federal government has a 98 percent stake in the profits, therefore making the Federal Reserve banks a new federal agency under the direct supervision of the Secretary of the Treasury makes sense. This action would create a truly just monetary system and return the power to coin and distribute real federal money back to the federal government as the founding fathers intended. Energy is the most flexible commodity. Energy combined with technology can be used to fill all basic needs. It may be possible to back every real federal dollar, including real federal dollars given in exchange for existing Federal Reserve notes backed by nothing, with 1 gallon of a new perfectly clean fuel. This new fuel would be a form of liquid hydrogen at room temperature that contains no carbon or other contaminates. The fuel would run as well or better than high-octane gasoline in most existing vehicles. Setting such an exchange rate would lock fuel prices down to 1 dollar per gallon. Before April of 2004 I will be able to determine whether or not this is can be done from a technical and practical point of view. Such an arrangement would also eliminate reliance on foreign oil and make the U.S. a major energy exporter helping correct the current account deficit. From a purely legal point of view only the federal government has the power to coin money. Today a private corporation called the Federal Reserve prints the currency in use. Just because the Federal Reserve makes a claim on the printed money saying that the paper is legal tender does not change the reality of the constitution. No constitutional amendment has been passed formally giving away the federal governments power of coining money to a private company called the Federal Reserve. The coins in circulation are real money, not the paper. If the Federal Reserve is not federalized people will continue to walk around with money that is backed by nothing and also unconstitutional. Eventually, this situation could result in the dollar becoming very weak around the world.

Personal success can become very difficult in an economic environment slanted towards the success of a very few established individuals. In comparison, it is relatively easy to reach out and educate a few people in order to change the whole environment for the better. If part of the education is that the people educated should turn around and teach others as well then education proceeds exponentially. It is in the best interests of the individual to care, and to act in the best interests of the majority. Some people believe that things would be great if they could just retire rich and do nothing for the rest of their lives. That goal is an illusion of happiness. In fact, that would be a very boring existence; most people would go stir crazy doing absolutely nothing for a very long time. Balance is much better. Spending time working on projects that one enjoys, or projects that one knows are needed by ones fellow citizens, makes the time off that much more enjoyable. Today the U.S. economy is being artificially supported by many foreign nations who are delivering over 400 billion dollars more real product to the U.S. than the U.S. is exporting. This has been an ongoing situation for many years that is only supported by faith in U.S. paper. That situation cannot continue forever. The economic situation can and should be fixed before the situation goes over a cliff. It is much easier to prevent hitting the bottom of a cliff if one takes action before actually walking off the cliff. The fact that the U.S. Supreme Court continues to uphold the Federal Reserve act of 1913 sheds more doubt on the competence or integrity of existing judges.

The reason why some jobs become tedious is because some supervisors fail to appreciate individuals for the job that they are doing. Those supervisors should be fired and sent back to work at the lowest levels until they learn humility and how to treat people with dignity and respect. The attitude of the supervisor, and team results are what is important. A really good supervisor manages a successful team with high morale. That does not mean a supervisor should never fire anyone. Sometimes people are just not ready to handle a particular job, and would do better at something else or with more education. In that case people can be fired with dignity and respect. There is also such a thing as tough love. People can tell if a supervisor is just putting people down out of meanness and arrogance, or if the supervisor ultimately cares about individuals and the success of the project.

Just like a company, a nation can own things. These things include everything that has been purchased or fought for since the birth of the nation. The U.S. is like a big rotary club with major club resources. The nation is run by elected politicians who can do whatever they want to with club resources, within certain guidelines set down in the constitution. If politicians decided to enact a law that violates the constitution, the job of the Supreme Court is to step in and say that is wrong and nullify the law. This is called checks and balances.

The president runs the executive branch. Since the executive branch represents a great deal of responsibility in the hands of one person, there are a many checks and balances that the executive branch is subject to. Supreme court judges are appointed for life. Judges cannot be removed by acts of congress or even by the president. The president can be impeached. Impeachment involves immediate removal from office by a two-thirds majority vote in congress. The president does not have the final say regarding details related to legislation or passing laws. That is left in the hands of congress. The president does sign the laws, and may veto packages of laws. The president does not have "line item" veto authority, which means that president cannot veto particular parts of legislation. Congress can over ride a presidential veto with a 2/3-majority vote. The president leads the military, but the president cannot go to war without the approval of congress. These are checks and balances related to the executive branch.

When it gets down to it, a president is only as effective as that president is supported by congress. Working together, the president and congress have great flexibility. Working apart very little is likely to be accomplished. This is how things should be in democratic nation.

The U.S. has plenty of checks and balances in place against both the executive branch and against congress. Both branches are subject to popular election on top of the other checks and balances I mentioned. There are no checks and balances against the judicial branch once a judge is appointed. The fact that a judge is appointed means that the judge is not even subject to a democratic process to begin with. That is putting a great deal of unlimited judicial power into the hands of just a handful of judges for life. That is the way the constitution intended it to be, and maybe it should stay that way. On the other hand, as more and more people realize that rejuvenation and eternal youth are now a reality, congress could consider at least one balance against the judicial branch.

Once the president puts a judge on the bench that judge answers to no one for all time. All other leadership posts are subject to regular validation by popular election. That makes the president's initial appointment very powerful. It is possible that a judge could become unable to function in office yet insist on staying in office against the opinion of 99 percent of the population including all the elected officials. This is an extreme example, but it is true. The judge could go insane or become ill and just not want to give up power. The constitution makes no provision for such possibilities, however remote those possibilities may be.

The reason why the constitution set things up like this regarding the judicial branch is because the founding fathers wanted to make sure that judges would not be affected by politics and would stay focused on preserving rights and the intent of the original constitution. The only problem with that argument is that it assumes that every judge appointed will be perfectly dedicated to preserving the original intent of the constitution. There is only one possible way to check the judicial branch. It is possible for congress to pass constitutional amendments. If enough people really felt strongly about a judge, believing that judge was abusing power and not really upholding the constitution, a constitutional amendment could be passed subjecting judges to impeachment by a 2/3-majority vote in congress. Judges are human like everyone else. The intent of a democracy is government by the people, so I do not see any reason why a judge should not have to answer to congress just as the president does. If people decide 200 years from now that judges should not be subject to impeachment such an amendment could always be repealed.

There is a one thing that I have been concerned about looking at the actions of the judicial system. The constitution did not create a Godless nation dedicated to atheism. I do not agree with taking God out of the pledge of allegiance. The separation of church and state is not about promoting atheism, or preventing voluntary public recognition of God. Granted, people may disagree among churches made by man. The government should not take sides in that debate. When the government takes the side of atheists, against all people of faith, taking God out of the pledge of allegiance for example, a nation has indeed gone down a very dark path. If anyone were looking for and abomination of abominations this kind of thing would certainly make the list. One insane individual walking into a temple and claiming to be God would be nothing compared to an entire nation turning to atheism. In the past entire nations have embraced atheism through the teachings of communism. That was an abomination. If the U.S. were to go down that same path it would truly be an abomination of abominations. Making a public display of the Ten Commandments illegal is one more step towards an atheistic nation. Eliminating the Ten Commandments is a victory for atheistic moral relativism arguing that God does not exist and that there is no such thing as true right and wrong. When a judge says that God does not exist, and that there is no right and wrong beyond what that individual judge decides, then that judge is in fact claiming to be God over the people of a nation arriving in the court of that judge. That attitude is the foundation of most evil done in this world. An entire nation accepting that attitude and writing it into the law of the land is a step in the wrong direction that needs to be apposed. It is within the power of each individual in a democracy to take up a cause and set things to rights through congress.

Advocating atheism is not what the founding fathers had in mind. The creation of individuals with divine or self-evident basic rights is an integral part of the constitution. I would appose a judge acting to change one nation under God into an atheistic state. Recognition of God on U.S. currency, and as part of the pledge of allegiance, is part of the constitution. If that isn't clear enough to certain judges in California or Washington, maybe a constitutional amendment is needed allowing impeachment of judges by congress to make it clear. People are free to worship as they see fit, and incorporate churches as they will. If a person decides not to participate in a pledge of allegiance that is fine. People should still have the freedom of public speech in addition to the right to remain silent. A handful of judges must not be allowed to take away that freedom of speech and expression from everyone without consulting voters. I would not require a judge to display the Ten Commandments, but at the same time I would not take that right away from an individual judge or a particular court district.

The previous paragraphs are an example of something I strongly believe in shaped into a political argument. That is what politics is all about. Elected officials are responsible for presenting ideas in a logical manner, fighting for the interests of the electorate. In congress, sometimes that involves compromise when it just isn't possible to get enough people to agree on a certain plan. It is always best to go into a debate strongly and without compromise, then only compromise if absolutely necessary. There should be no compromise regarding keeping God in the pledge of allegiance, and allowing teachers in our public school system to lead students in that pledge who choose to participate.

When it gets down to it the real power is in the hands of the electorate. People are not going to keep someone in office whose actions go against the ideals of the electorate. The nation will inevitably take the path of the majority. The public elects every congressional leader plus the president. Not only that, over time the public pulls people out of office and replaces them with individuals whom they believe will better represent public interests. Things are as they are because that is the way most people want things to be. Until people really understand that, and realize that political battles are fought in coffee shops not in congress, things will continue much as they have been. An object in motion tends to remain in motion unless acted upon by an external force. In politics that can translate into—if the electorate is confused or has no particular plan then things will not change.

People who are elected honestly try and represent the people who helped them gain office. Some times special interest groups appear to have more influence over a candidate that an average voter. Did a bunch of average people make contributions to that politicians campaign? If not, can we blame the politician for being influence by particular groups? The truth is that if the majority really cared who was put in office, there is no amount of money that could be put in by a small group to sway a race. If 150 million people all put in a couple dollars, which everyone can afford, that would be more than twice as much as anyone has ever put into a political campaign. Not only that, if 150 million people decided on a particular candidate all they would have to do is vote.

Special interest campaigns only work to the degree that people allow themselves to be influenced by those campaigns, and to the degree that the public really doesn't care much about the election anyway.

Politics is a win all lose all game. Fifty-one votes beats 49 any day of the week. Forty-nine people may be adamantly against a particular campaign platform, control of 100 percent of the club resources still goes to the 51 with the majority. This sounds unfair, but it is best we can do in a real world situation. Decisions are black and white. For example, we could not build 40 percent of a new shuttle if 40 percent of the people apposed building new shuttles. At the end of the day a nation needs decisive leadership. Without decisive action there would be chaos not leadership. A democratic system is the best of all possible systems. Democracy allows for an open forum debate combined with the decisive action that is needed to run a nation.

Public resources include the military, public land, and the gold in Fort Knox. Public resources also include roughly 20 percent of Gross National Product. That is about what we end up collecting in taxes. In a presidential campaign one major party will waste $100 million dollars not being elected. Granted, sometimes the winner will make concessions to strengthen the party position by including apposing party views. This isn't a requirement, and probably does not happen often. A political donation is like a bet not an investment; you either win or you lose.

Two hundreds years ago grass roots politics was all there was. Mass communication and media conglomerates did not exist. Grass roots politics is still the best political campaign.

Mass communication techniques are not that effective compared to personal communication. When was the last time you changed your mind and took advice based on something you read in the newspaper? Probably not as often as you have taken the advice of a close friend.

Even though mass communication is not very effective, it is better than nothing. If nobody around you has any particular opinion that they feel strongly about, the only thing that you could be influenced by would be what you are exposed to in the press. Often, we know very little about candidates at a personal level. An open public debate is a good way to get to know candidates. Debates are spontaneous. Often public debates reveal what the press might hide.

The ideas in this book are essentially a political platform. This platform and plan could be used by anyone. Usually, politicians present very broad and often vague platforms. That is probably because they do not want to offend anyone, and hope to win by getting a little more name recognition via the media. That is not true leadership, though I am sure that method has won more than one election.

Leadership involves setting very particular goals and then achieving those goals. People are capable of a higher level of understanding than today's political campaigns would seem to give people credit for. There is no reason for not presenting specific goals and detailed plans going into a national race. It takes more than good intentions to produce positive change and create a brighter future for an entire nation.

How do we know that the politicians we elect are capable of setting the right goals and making it happen? It certainly would not hurt to have some kind of a detailed written proposal from individuals seeking public office. Granted, there is no requirement for such a thing in our current political system.

Two hundred years ago when people went out campaigning it was a very personal experience. Party leaders were familiar with a detailed political platform that the party was running on. Those individuals would visit towns and cities, riding in on horses. Those visits made up the majority of the campaign effort. Party members would ride into town and jump up on stumps to address a crowd. They would talk about the party platform, answering people's questions, and dealing with hecklers. Of course, the other party had supporters to, and they were not shy about voicing their opinion publicly. Some hecklers were paid to heckle regardless of what the other party representative said. Politics was risky business in those days. People on both sides put their heart in to it. Too many hecklers and a speaker might be tarred and feathered, led out of town backwards on a donkey.

The purpose of this book is to empower people with knowledge and solutions. These are the tools of national politics down at the coffee shop level. The level of interest in politics rises and falls based on the importance of issues and how strongly people feel about them. If the world faced no new challenge on any front, continuing much as we have in the past would be no major problem. Facing no immediate challenge, there would be little incentive to rush towards positive change. It is the nature of most people to close the barn door after the first horse is gone, or not rock the boat so to speak unless there is a real concern that the boat could be sinking.

I believe strongly in destiny. I have seen it at work in my life and the lives of other people. I believe that from the viewpoint of an all powerful creator what is most important is the development and progress of society in general. The trials and tribulations of a single individual, during a single lifetime, are less important. Many people site the existence of crimes committed against the body as a reason to doubt the existence of God. First of all, if God were to intervene against every physical injustice free will would not exist. More importantly, clearly God would be more concerned with the spiritual side of life than the physical body, even though the physical body is still important. Until an individual recognizes the reality of spiritual existence, that individual will know nothing of God. Righting physical abuses or injustice is as much or more the responsibility of mankind as it is the responsibility of God.

A clear line exists between political justice, which is represented by justice delivered by governments however right or wrong that may be, and the judgment of God. There is good reason for individuals to have a certain respect and fear of God. Divine justice may appear nonexistent at times, or slow in coming, but there is good reason for that. One life is like a drop in the ocean of time looking at the spirit. Most Christians do not speak of reincarnation. Achieving the level of spiritual awareness necessary to be one with God and exist in a purely spiritual sense could be called heaven. Failing to achieve that, people end up getting themselves reincarnated, kind of like someone finding their way back home in a drunken stupor not knowing how they arrived, only that things have not changed much. That would be a low level of spiritual awareness after death, and could be called purgatory of sorts, with the possibility of waking up spiritually and doing better in a new life. Hell would be an extreme situation representing the end of the line, or the point at which an individual spirit no longer has a chance to play an active role in this universe or even a chance to try to do better next time. An individual spirit is very real, and does have a location and energy characteristics in this universe, assuming that the individual spirit is not capable of recognizing God and becoming one with the entire universe now and then at will. An individual spirit without that ability can be cast into the core of a heavenly body and be trapped for eternity experiencing pain and contributing some energy to the building of planets or stars. The Lord is slow to anger because the consequences of unleashing that anger are so extreme. I hope that clarifies the need for individuals to focus on fair political justice rather than blaming God for everything that goes wrong in the world. That attitude of blaming God rather than fixing that which is the responsibility of individuals to fix, via political activism, is almost as bad as committing the actual crime. So, it is best to leave judgment in God's hands and not ask for more of it to be delivered unto others. Freedom, both physical and spiritual, is a very valuable and should never be taken for granted.

As technology progresses it becomes more difficult to avoid greater access to weapons of mass destruction. Weapons of mass destruction may even become more destructive. In that case the only solution will be improving every individual such that no one goes off on a massively destructive path. Certainly intelligence services can help, but that is just a temporary solution not a cure. Actually changing the hearts and minds of a large number of individuals is a project that many philosophers have considered impossible. The reason why people have considered this impossible is because there isn't much of a precedent for it looking at history. Even so, it is possible. The way to make it happen includes education combined with a massive grass roots political campaign.

Here are a couple ideas for establishing a grass roots political campaign. This campaign would involve anyone who wants to run for office on a platform similar to what I discuss in this book.

The first idea is to put the party back in the term Political Party. There is a reason why we say Political "Party". A long time ago people would host parties at their house. The whole political network centered on throwing parties and entertaining guests. Those house parties were the birthplace of new political platforms and plans for the future. Throwing parties is a great way to make new friends and meet friends of friends. Today's parties often focus on how much alcohol can be consumed. A long time ago people had a broader range of concerns. Certainly beer, wine, and good food were a big part of the party. At the same time, people talked about the future, and the goals and dreams they had for a better life. These conversations resulted in political campaigns and scheduling political activities some time after the party.

There is nothing closer to grass roots than doing door-to-door canvassing, surveys, etc. The marketing text on the back cover of this book was survey going door-to-door to gain an understanding of how people would react to the book. Meeting people and learning how to best package the book so that as many people as possible would read it after seeing it in a bookstore was important. Without a friend recommending a book, the only thing people have to go on making a purchase decision is the cover of a book and the marketing text on the back. It is not possible to interest everyone no matter how good a book cover is. Surveying hundreds of people and asking if they would or would not buy the book judging by the cover, then if not why not, makes it possible to interest a higher percentage of people. The same is true about politics, since politics is actually marketing ideas. There are many different ways to present the same idea. Talking to many different people it is possible to determine what presentations work best. A really good politician is able to judge the person in front of them and select a particular presentation that is best suited for a particular individual. Of course that cannot be done with a book cover, since the cover is what is. That is why a personal approach is always the best marketing approach.

If this book is to be used as a political platform, the availability of the book is important. If this book is not available in a bookstore, just purchasing one book from the store could result in the book being placed on the shelf where hundreds of people will have the chance to see it and purchase it. In addition to bookstore retail outlets, this book will be available at wholesale prices to anyone purchasing ten or more copies. This book is not difficult to sell, there isn't anything even remotely like it on this planet. So, doing a grass roots door-to-door reach out, people could make some extra money selling copies of the book.

Here is some advice from selling a few copies of my own books going door-to-door. Many people are suspicious or afraid of strange people knocking on their door. Leave some space between yourself and the door, then when someone answers the door hold up the book to the left or right and admire if for a moment such that the person can see yourself and the book. Do this rather than talking to the person right away. Most people will be drawn out of the house and towards the book. Say something to the book like "I have a new book" or whatever, the choice of words is much less important than the admiration. Most people will then be curious about the book and want to buy it after asking some questions.

Running for office, be it a national office or local office, is a challenge. I would like to help other people who decide to run for office using a platform that supports what I am trying to achieve. I will do my best in that respect. If more than one person from the same area would like to run for the same office I would encourage people to talk with each other and see if an agreement can be made to support one or the other. A party primary is the election in many cases. It is a similar situation having two people running on the same platform whether they be from the same party or independent. There is no set limit to the number of primary candidates, however running many different primary candidates spreads out party resources that could otherwise have been used to promote one strong candidate in a geographic region. This is one reason why the Democratic Party will lose the presidential election in 2004. The Republican Party is united behind president Bush while the Democratic Party will remain divided until the Democratic primaries are complete. The year spent campaigning during a primary, or slandering other people from ones own party, is an important year for a political party. When it gets to the general election there will only be a couple major party candidates to choose from. Clearly, the primary is at least as important as the general election, since the primary narrows the field from a potentially unlimited number of candidates in primaries down to two major party candidates and some minor parties or independents.

A party primary is a very expensive race if one is serious about winning. Considering the resources that go into primaries, and that those resources could be used in the general election, a party primary is something to think about. If two people are from the same party and agree on most things it doesn't make much sense to battle each other. The Republican Party has done a very good job of uniting behind candidates. In my opinion Republicans tend to get along with each other and support each other more so than people in the Democratic Party.

I will run for President of the United States in 2004 as an independent candidate with a strong Republican bias. In congress, one major party needs to become dominant to avoid filibustering and allow for congress to get the job done more quickly. Lengthy debate is a luxury for people that are not faced with any significant near term challenge. I will run as an independent in 2004 because I know exactly what needs to be done for this nation to survive and prosper into the future, and because I believe I can win. Looking at all possibilities, a Republican might ask what if I did not win and managed to hurt the chances of President Bush by pulling Republican votes. Although I prefer the Republican Party, given the situation as I see it the end result for this nation will not be much different on the path chosen by existing Republicans or Democrats. Today, neither party plans to mobilize this nation to prepare for the future, though people may change their mind in that respect after reading this book. Both major parties are strongly influenced by special interest groups. At this point the reality is that the two major parties represent two different fronts of the same special interest group giving people the illusion of choice while things stay the same for the most part whatever candidate is elected. So, if I lost the race I would be more concerned about the future of the nation than I would be about which major party held the White House. There are differences between the two major parties, but these differences are superficial looking at the broad social picture. In the past neither party has tried to change the Federal Reserve System nor have they tried to change the form of society from centralized systems to decentralized systems.

Winning a presidential election has always meant winning a major party primary. That is obvious considering that it generally takes a majority in the popular election to win the presidency, and a minor party by definition does not have a majority. Hopefully I can overcome that reality even if it has never been done before. Since major party primaries are the norm for winning elections it is a good idea to consider what it takes to win a primary.

Most of the time the person with the greatest primary campaign budget wins the primary. Name recognition is also very important. The people who lead the major parties behind the scenes back certain candidates in the primary election. Communicating with hundreds of millions of people to win an election is always going to involve money. Anyone working full time on a campaign is going to need a paycheck. Considering these basic realities, special interest groups with large budgets have dominated the political scene in the past by selecting certain candidates. Recently, a new law was passed to make campaign donations reflect public support by limiting the amount any one individual or corporation can donate. This is a new law, so it is to early to say how this will affect the political process. The limit on individual donations to candidates is 1000 dollars. Minor parties and independents are excluded from the law. The law was passed to limit the influence of special interests groups on major parties. It is clear that a special interest group would not waste money on a minority party that has never won an election. The fact that this law actually passed congress indicates that politicians must have felt so bad about the influence of certain special interest groups that they were all willing to refuse that money entirely. One can only imagine how bad things must have become for politicians from both parties to have had an ethical epiphany and turn down money now and into the future. Theoretically, as the law is now written, each individual or corporation should have no more than 1000 dollars worth of influence on any particular campaign. This law has the potential of putting enormous power back into the hands of average people. That completes a summary of the current national political scene.

Consider the international political scene. The executive branch leads the charge when it comes to international politics. At first glance, it would seem that the U.S. has enjoyed very little peace in the past 200 years. The men and woman of the armed forces have allowed peace to reign on our home ground, and that is very important. Sometimes freedom has to be fought for. The defense of freedom in foreign lands does not mean the U.S. is a nation bent on conquest and subjugation of foreign lands. The U.S. has conquered in the defense of freedom. That intent is important. It does not mean that the United States has always been right; the intent means that the U.S. has been trying to be right. In both Germany and Japan the U.S. rebuilt those nations after World War II for the sake of the people that were led astray. Those benevolent actions were unique in history. Only a Christian nation would even consider forgiveness on such a broad scale. Any other nation would have kept those conquered lands and exacted tribute, certainly that is what Germany would have done if Germany had won the war against the U.S. The U.S. did not do that because it was not the right thing to do. Imposing slavery leads to a degradation of the very heart and soul of then enslaver.

Today the world is a different place. During World War I and World War II the U.S. faced organized military machines bent on global domination. Today we face different threats, each with their own agenda. Things are greatly complicated by weapons of mass destruction. I expect things to get more chaotic around the world before they get better. Whatever challenges this nation faces we can count on the strength and intelligence of the men and women of the armed forces if diplomacy fails. This nation will remain free.

President Bush is making great progress in the war on terror, and building up national defense including the deployment of a missile defense system. That effort should continue, if not accelerate. At the same time, we can do a better job reaching out to other nations. That is not an easy task, but is an important task that can pay great dividends and it is one of those things that can always be improved.

There has been talk of an axis of evil in this world. Sometimes people ask why would God allow evil to exist? I believe the answer to that is obvious. Free will, and the idea that individuals should act as good stewards, trumps direct divine intervention whenever possible. The only exception would be allowing the annihilation of everything. In that case I would expect just enough divine intervention to get past the crisis. God has a patience that spans lifetimes, and a capacity for forgiveness that is greater than individuals generally extend. If everyone acting in an evil manner in the eyes of God instantly went up in smoke and eternal damnation that would limit free will. It would also limit the concept of forgiveness.

Identifying an entire nation as evil can be misleading. It is true that a nation can be led by relatively evil individuals, and that the people of those nations are partly responsible for bringing certain leaders to power. No one is completely evil, even if some individuals appear to be striving for that appellation. Our intelligence community looks at these things very closely, recognizing all the shades of gray that are involved. The U.S. intelligence community itself is certainly no saint. When it gets down to it people have to learn by their mistakes.

Can anyone honestly say they know exactly what is right and wrong in every situation in the eyes of God? For example, one of the Ten Commandments says thou shall not kill. Does that mean that everyone who has defended their country in the armed services is evil? There is a difference between a static moral code, moral conduct, and moral relativism. A list of the right thing to do given any particular situation in this universe would fill a trillion, trillion, trillion books. Just listing out all the factors involved in any one situation would take a very long time. The Ten Commandments represent guidance towards the right path. There are some situations that require individual judgment. For example, if someone built a home made nuclear bomb and walked off to detonate that bomb in New York, would it be wrong to steal that bomb from the individual? Of course not. Moral conduct exists in the light of a higher level of understanding embracing the intent of the Ten Commandments. Thou shall not kill is intended to prevent murder that is not committed in self-defense. Thou shall not steal is intended to allow the ownership of private property. If an individual is intent on using some private property to murder millions of people moral conduct in the light of understanding allows for the confiscation of that property. Moral relativism is a philosophy stating that there is no right or wrong. Moral relativism proceeds logically from atheism. Moral relativism is wrong.

When an individual or nation acts to bring a greater freedom and justice into this world, that action is taken without tearing apart the very fabric of free will in this universe. The banishment of evil by divine intervention is not an ideal solution. Students becoming teachers is a much better solution.

Evil is as much ignorance as it is anything else. Evil people tend to act in a way that limits their own life, liberty, happiness, and the happiness of everyone around them to some degree. Evil often associates itself with cowardice, expecting people to sacrifice principle in the face of a threat or actions of violence.

Can anyone doubt the ignorance of Saddam Hussein? The man failed in his first assassination attempt to gain power. The reason he ended up in office was by appointment from his cousin who gained power. Since then, Saddam has miscalculated every military action he has under taken. The only reason he wasn't completely over run in the Iran–Iraq war was his decision to launch massive chemical weapons attacks that killed almost as many of his own troops as Iranian troops. He turned his own local Kurdish population against his regime by attacking and killing many thousands of Kurdish civilians with chemical weapons. Foolishly, he invaded Kuwait to be thrown out in a matter of days by coalition forces. Saddam Hussein decided that a dictatorship would be better than a democracy.

Iraq is just one issue on the stage of international politics. North Korea is another example of relatively ignorant and cruel leadership. North Korea is falling apart. How can people live and work in peace when they are threatened with being dragged off to the Gulag for uttering a word of protest against a leader that they did not choose? Freedom of speech is very important, as important as the right to bear arms and elect leadership. The dictatorship of North Korea is actually worse than the dictatorship that once ruled Iraq. North Korea is a communist dictatorship. Communism translates into no freedom of religion, plus every hour of work a person does goes to the government.

Communism is an economic system based on the theory that everyone must give everything they create to the government and hope they get enough back to feed the family. When people are allowed to own what they work for life is much better, and the net sum of all products produced is greater. That makes it easier to care for the few people who truly need a hand up to provide for themselves. The other problem with communism is that it was originally intended to be atheistic and dictatorial. Softer forms of communism exist. That just goes to show that even people who lean towards communistic slavery will only tolerate atheism, dictatorships, and destroying individual freedom to a certain degree. I studied communism as part of my philosophy minor at WPI, and received an A. Complete understanding of the ideas of a group can be the beginning of change. Sometimes old systems have to fall to build something better. Completely understanding an adversary is the best way to win, since that understanding brings about an ability to predict future actions.

Political goals, destiny, and individual actions all play a role in shaping the future on this planet. Many futurists try to predict the future. Prophecy is an integral part of many religions. Even ancient cultures like the Mayans wrote prophesy extending to the present day. We know this because the Mayans were extremely good with calendars. So, what will the future hold?

12

Future

The future is difficult to predict because the future represents the combined actions of billions of people in a changing environment. Each one of those people contributing to the future has free will. Fortunately, forces are at work greater than any one individual making it possible to predict future trends.

Each individual is faced with dozens of competing interests at any moment. Which way will they go? That is exactly what you would have to know to predict the future exactly. It is true that the number of people contributing to the future is a huge number. It is also true that the rate at which an idea can spread is a huge number, so that evens things out. The math of three people, telling three people, on and on can result in a number up in the tens of millions very quickly. That is not even considering mass communication that is able to instantly reach millions of people. Ideas really do shape the details of the future, so it is important to recognize how fast ideas can move trying to predict the future.

It is possible to get an idea of people's intentions by observing their actions. It is also possible to recognize practical aspects of broad situations that effect many people. Predicting the course of the future involves recognizing the path that people will take in a given situation. The key to influencing other people is truly understanding other people. It is important what is truly in people's heart. Sometimes people put up a front, or are reluctant to express what they truly believe. If an individual is presenting a less than honest façade, that facade has to be pushed aside to communicate an idea. The way to do that is to understand the intentions of an individual, and gain enough trust to really talk to someone. Intentions are reflected in actions. Intentions are also reflected in how people say things more so that what people say.

There are two broad categories to look at predicting the future. The first category involves the flow of new ideas introduced into society, combined with existing beliefs, all contributing to individual action. The second category of future prediction involves external forces that influences society. Both internal and external forces need to be considered to predict the future.

Understanding both the environment, and how large groups of people can be expected to react considering the beliefs that are published by those groups, it is possible to make very likely predictions for an entire civilization. That is only possible because large groups do exist with similar beliefs. If not, prediction of the future would be nearly impossible.

Consider the future of this planet given the environmental changes during a pole shift, and in the context of how you would expect your friends and enemies alike to react. That is future 101.

Religion, or a lack of it, does unite people from different nations. The U.S., United Kingdom, and Australia hold very similar religious beliefs and values. I would include Brazil in that group as well, since Brazil has strong Roman Catholic population and quietly influences South America. Brazil is not an English speaking country though, and so Brazil is not as close to the U.S. as England and Australia. Most of the Middle East favors Islam, with the exception of Israel, which is sort of a wild card. Christianity is also the major religion in Europe, but attitudes are a little different in Europe, and so Europe cannot always be expected to follow the lead of the U.S. Europe is also represented by a diversity of nations and languages, so it needs to be looked at separately. Since global environmental change is going to factor into this analysis of the future, location is very important. All it would take is a 100-foot rise in ocean levels to submerge almost half or Europe. That is just a rough guess, whatever the exact number is Europe is in danger losing a majority of land mass. The situation is much better in the U.S.

Russia and China also represent major players in the future. Even though both of these countries use to be atheist communistic nations, things are changing rapidly. Christianity is spreading rapidly in China. Christianity is also the most popular religion is Russia. Russia and China are slowly moving towards free market economies much like the U.S.

I have not mentioned Africa. Africa is not a global economic or military power. People may flee to Africa since the continent of Africa is at a relatively high elevation and not on major fault lines. That observation is positive only from a relative point of view. Africa could still break in two or three pieces, but at least the pieces would remain intact and above the ocean. Not being a global military or economic power, Africa will not have a significant near term influence on the future. Any analysis of the future has to be limited to some scope and level of detail or it would become impossibly complex. So, the scope of this analysis will be major international players.

That breaks the world down into a manageable number of groups to consider. Today the U.S. is taking on the role of global mediator and sometimes-global police officer. As the influence of the U.S. grows, along with the number of nations who are allied with the U.S., the U.S. will respond to broader threats. The U.S. waited to get involve in World War II until Germany was on the verge of taking Europe. America turned away from the realities of the holocaust occurring in Germany, favoring isolationism for some years. The attack on Pearl Harbor changed all that. Attitudes are different in the U.S. today compared to the isolationist trend that existed prior to World War II.

Freedom and democracy have spread across the world in a way that would have made the ancient Greeks proud. Although conflict seams to be wide spread, since every item is covered in the mass media, it has been worse in the past. Mankind is not faced with actual World War, which did happen twice in the last century. The U.S. is reasonably united in purpose today, though not as much so as during World War II.

Beginning with acts of terror on U.S. soil, a massive effort has been launched to focus on preventing future acts of terror. That effort is going very well. Many unsung heroes should be credited for the relative safety of American citizens after 9/11. Intelligence services do not disclose successful operations, so no public credit will ever be given. It is only intelligence failures that make it into the public eye. That is just the way it has to be. The relative safety of all American citizens since 9/11 speaks volumes for the efforts of President Bush and his team. Whatever the future may hold, chaotic terrorists must not be allowed to rule the day. A peaceful work environment is necessary to make forward progress.

Now that the United States has taken center stage in a world of hopes, dreams, fears, and tears what will the nation do with the responsibility that power and knowledge convey. Power and knowledge do convey responsibility. Responsibility can be defined as the ability to respond. An individual, who knows that something wrong is being done, and who knows that the power to do something about it is in their hands, will feel a certain degree of responsibility allowing that bad situation to continue. The relationship between knowledge, power, and responsibility is simply human nature. Most decent people do try to improve things if they know what needs to be done and they have the ability to make it happen. Understanding this part of human nature helps predict the future. In fact, I'm counting on this being true. This book in an effort to increase knowledge and convince people that they each have power to change things especially when joined together in common cause with other like-minded people. So, what will the U.S. do with the influence the nation now has on the global scene? I expect the U.S. will bring peace if possible, and freedom from tyranny when necessary. The U.S. will teach people how to provide for themselves. The U.S. will help bring democracy, freedom, and basic human rights to those that want these things, and are willing to stand for them. These are American values. It is not possible to help someone that doesn't want any help, and to much help can also be a problem creating a never ending dependency.

Lets begin with some specific predictions, and reasons why these things are likely to come to pass. I expect an economic boom in the U.S. starting in 2004. Before then, it is possible that the economy will weaken for a time. The problem with the economy right now is simply a lack of effective leadership. Far to little is being done, and more importantly the actually causes of economic stagnation are being ignored. The truth is quite simple. GNP is a measurement of products being produced. Are people going to keep producing more products if they don't think they need many more new products? Of course they won't. How many television sets do we need in the U.S? How fast is fast enough for a personal computer? That point has probably been passed some time ago. How many cars can one person drive? That is an easy question. They can drive one car, and I imagine that more than one serviceable car now exists in the U.S. for each person with a license. How many houses including apartments are needed to house everyone in the U.S? Without doing a formal survey, just considering vacancies, I would say that this nation probably has built close to enough typical wooden above ground housing units for everyone. Granted, we can always keep improving the quality of housing, building mansions for everyone if that is what it takes to get the economy going. These realities are being ignored. Until people recognize what is actually happening the economy will just flounder along at whatever it pace it feels like almost like a roll of the dice. My prediction is that at some point after the release of this book enough people will realize that new products really are needed and then people will get busy building those new products I mentioned in the survival system chapter. That would create an economic boom.

One danger of making predictions is that sometimes they can be self fulfilling. For example, if after reading this book many people decide not to get a new car, computer, or house that could bring about the economic dip that I mentioned. It is important to keep the assembly lines going. The air cars I mentioned would be built using the same assembly line technology that GM is using right now, including many of the same machine tools. The solution would be to quickly start producing new technology in volume such that the rate of new production out paces the decline of older technology.

Even faced with the reality of an extreme environment it will be possible to make progress by creating manufacturing facilities able to operate in that environment in addition to survival shelters. I am confident that the U.S. will make the necessary preparations to carry civilization through very difficult times. My confidence is based on several trends.

The majority of people in the U.S. are hard working, intelligent, Christian people. I do respect people of all faiths and religions. Looking at the future economy, the easiest way to predict the future is predicting the course of the majority. Christianity is the majority religion in the U.S. and the beliefs of Christian will drive the future of this democratic nation. If the majority of people in the U.S. were Muslim things would be different, but that is not the case. It takes a certain degree of Christian faith to take earnest action along the lines I have described building survival systems. Other religions may take action based on the purely technical aspects of what I present. Christians will have both a technical reason plus reasons found in scripture. The near term future environment that I predict is not unlike the events of Revelation.

Islam does not include a book of Revelation. More importantly, very few people in the Middle East have the technical background necessary to understand the models that I present in this book. Also, this book is written in English. That means that this book is not likely to influence many people in the Middle East. People in that part of the world can be expected to go on doing pretty much what they have been doing until some cataclysm moves them to desperate action.

I expect the publication of this book to rekindle faith in the U.S. and unite Christian people around the country. So many people and generations have held on to the faith 2000 years after the birth of Christ. Today Christians study the bible, and many people try to interpret scripture and live better lives. I expect the interpretations that I have provided to become popular since these interpretations are back up by hard science and products. It will take some time for really convincing rejuvenation examples to spring to life. Being 35 and looking 20 in all respects is not very convincing, though I am personally thrilled since I did age up to 34. There are a few people in Hollywood who take very good care of themselves and look quite young at 40. In time, some well known public figure in their 40's or 50's is going to rejuvenate in front of peoples eyes over a period of months back to their early 20's. That will convince just about everyone in the nation.

Europeans tend to be more liberal in some ways than the average person in America. This is not true in all cases. I am looking at general trends to try and better predict the future. "In God We Trust" is not written on European currency. I expect Europe to adapt slower to change than the U.S. for several reasons.

Germany is sold on the idea that mankind has caused a global warming crises by burning fossil fuels. Very little of the change we see today is driven by fossil fuels, beyond the immediate layer of smog over some cities. The Green party in Germany won many national positions. I expect the leadership in Germany to go down a wait and see political path rather than taking preventative measures.

The attitude that mankind is responsible for global environmental change, rather than this being mostly a natural cycle, results in different solutions. Had the U.S. elected Al Gore in 2000 the U.S. may have taken that same viewpoint. Global warming was a key part of Al Gore's campaign. Once people get things into their head it is very difficult to change their opinion. If this nation had started the ball rolling to handle global warming based on the assumption that it is was all the result of man made pollution, then people would be less likely to a fresh look at the situation. The reason is that pushing billions of dollars towards handling a particular problem, even if it be an invented problem, creates a vested interest in the problem. In other words, all the people employed to implement a particular solution to the invented problem will argue day and night without listening to reason just to keep their job. Germany has already gone done the path of fixing a man made global warming problem.

That is how I view the different approaches to environmental change, and how I see different countries reacting. Religion could also play a role in the decisions that people make. Have you ever become involved in big arguments with someone only to end up becoming the best of friends latter? There is one religious card that remains to be played. Germany has been getting into big arguments with the Church of Scientology harassing Scientologists and trying to prevent the rapid expansion of Scientology in Germany. In turn, this has only upset Scientologists more and caused Scientology to focusing many millions of dollars and much effort to counter that German attack on religious freedom. The net result could be many people in Germany agreeing with Scientology in the end. Do you waste time arguing with someone who you do not have to argue with, and whom you know is completely off base? Of course not, you would simply ignore such people. That is not what the German government is doing, and that could have something to do with why Scientology is in fact the fasting growing religion on the planet.

I visited Germany myself to participate in a march for religious freedom. Scientology is quite interesting. In the absence of the information presented in this book, Scientology makes a lot of sense. Knowing more about this universe and creation, gaping holes are obvious in the logic of Scientology, and in the attempts at spiritual advance that do exist in Scientology. The important point for this analysis is that if a nation like Germany leaned towards Scientology rather Christianity some interesting things could happen.

Scientology does not preach the idea of one God, or one creator of this universe. In one taped lecture L. Ron Hubbard goes as far as to belittle people who would pray to one God, saying that the individual themselves is all one will ever know of God for a very long time. That one tape doesn't represent Scientology's formal position regarding God. The formal position is that Scientology does not intrude into the "8th dynamic" which is called the God dynamic. Scientology considers that the individual has 8 dynamics, or 8 different urges to survive for the benefit of self, family, social groups, mankind, the animal kingdom, the universe, the spirit, and the Supreme Being. There is nothing particularly wrong with that statement. The trouble is that Scientology fails to recognize the active role that God does play in this universe, fails to understand the exact nature of creation in this universe, and in that light proposes solutions that either do not work or do not make sense considering the reality of existing circumstances.

Scientology does contain enormous amounts of information, much of which is true. The problem is that building the subject on a weak understanding of the universe results in false information and ideas mixed in with some very good information. In my opinion, and everything I have mentioned about Scientology is my opinion, the methods of spiritual advance used by Scientologists suffer most from the failure to formally recognize God and the truth regarding spiritual creation. In other words, there is a great deal of true information regarding individual psychology in Scientology, but I believe the chances are very low if not zero that an individual will achieve true spiritual freedom and spiritual ability relying on Scientology.

Just as Christians come up with some odd interpretations of scripture, Scientologists are no less prone to making mistakes. Telling people that they have 8 dynamics, and that the 8th dynamic is the God dynamic which can be reached by advancing through Scientology levels, some people are bound to start thinking they are indeed God. There is difference between theism and having a bunch of people running around thinking they are God. I'm not sure which is worse; atheism or thinking that one is in fact the Supreme Being to the degree that a Supreme Being does not exist beyond ones self. In any case, if the Biblical prophecy about someone claiming to be God in the rebuilt temple does in fact come true, I would not be surprised if the individual making that claim is a deluded Scientologist.

Scientology has the potential of fulfilling more than one Biblical prophecy. L Ron Hubbard mentioned that he would return after death and take Scientologists off to "Target II", earth being "Target I" of course. I can only assume that he intended to use spacecraft to accomplish this relocation. If Scientology was a small group in a one horse town, these things would not be significant. Scientology is the fastest growing religion on the planet, and it will influence the future in some way. I do not doubt that L. Ron Hubbard could be reincarnated, nor do I deny the possibility of civilizations existing in this galaxy with spacecraft technology. Only time can tell what the future will actually hold as far as all that goes.

It is tempting to discard these things out of hand, and say UFO's couldn't possibly exist. Personally, knowing these craft can be built, I tend to be more cautious. Reincarnation is possible, as is space travel, and it would be naïve to assume that mankind is alone in this galaxy at least until the future plays itself out.

Christianity also contains prophecy regarding reincarnation, and things like the rapture, so Christians of all people should not belittle these Scientology ideas.

Christianity looks to prophecy for the future. Scientology has no prophecy, not acknowledging an all-knowing God. One Christian prophecy says that the temple will be rebuilt in Israel, and that a man will claim to be God in the temple. It is hard to believe that a Scientologist would be foolish enough to do something like claiming to be God publicly in the rebuilt temple. Before any of that could happen, the temple would need to be rebuilt. It does like some kind of peace settlement is going to be achieved between Israel and Palestine. After that happens, I expect that the temple will be rebuilt.

Looking at things from a Scientologist point of view, it would not be unreasonable for a Scientologist to believe that they are God. The irony is that there is some truth in that. Having been created in the image of God, people can transcend the ordinary and at least temporarily be part of the mind of God. In fact, people cannot avoid being a part of God, knowingly or unknowingly. The error in Scientology is failing to recognize a real Supreme Being that is distinct from the individual. An individual can become part of God once again, having moved away in the beginning. That just means the individual has the potential of perceiving creation, not necessarily the potential of creating a new galaxy or controlling the oscillations of a planet.

Scientology philosophy is extremely complex, and very different from Christianity. Considering that hundreds of Scientology books exist, and thousands of Scientology tapes, it would be difficult for anyone to get a complete grasp on the subject much less a majority of the people who become Scientologists. If someone really understood the subject they would realize that L. Ron Hubbard never really figured anything out along the lines of God, but that he leaned towards the idea of the supremacy of man rather than the supremacy of a single God who created the universe and all the life in this universe. A very intelligent, or honest Scientologist, would recognize that these things were the opinion of L. Ron Hubbard and that it was necessary for individuals to do their own research and make their own decisions.

Scientology misses the fact that a range of spiritual power does exist in this universe, and that spirits or angels do play an active role. God is completely powerful and able to use that power at any time. That much power is just not the kind of power that a man can own and wield. The mere existence of that much energy on the planet would end the planet. God must work through lesser beings be they individuals, spirits, or even planet building spirits working on longer time spans. God created a universe of life. This entire universe from atoms up to galaxies represents forms of visible life. Besides that visible life, another 90 percent of the universe is invisible spiritual life with a similar range of size and power. The spirit that is an individual man or women is just one particular order of magnitude created to build civilizations. God can influence the very planet we walk on at will. Ignoring these truths could result in Scientologists doing some very foolish things.

The belief in multiple Gods, or one God for everyone, is one reason why Scientologists would doubt the technical information in this book regardless of how well it is proven in a lab. According to L. Ron Hubbard this physical universe was created by an agreement among the spirits of individual men and women, and the spirit of man is the top of the ladder potential wise. In other words most Scientologists do not believe in guardian angels working for a creator and influencing mankind. This gets back to a basic misunderstanding and a certain blindness LRH had regarding spiritual existence greater than himself.

L. Ron Hubbard says this is a universe created by agreement. I say this is a universe created by one God. Obviously we disagree, yet the universe still exists. I win that argument. L. Ron Hubbard was a science fiction writer. I have read some good science fiction lately presenting the idea that a large group of people could join together becoming more powerful and then doing neat things like taking a planet out of this universe into a new dimension. That is science fiction. It does not work, but it is an interesting concept. Electrons are extremely simple, and are able to merge together smoothly and create larger singularities of greater power. One might consider that this would also be possible for individual souls. The individual spirit is much more complex and does not merge together completely as do electrons. So, there is no danger of a bunch of Scientologists merging together and vanishing the planet into a very strong singularity. Even if this merger were a possibility, the combined power of a billion Scientologists would not accomplish such a thing on any large scale. Spiritual advance to levels of higher awareness and power has to be a personal thing, not a group agreement thing.

One would think that a subject called Scientology would have something to do with science or engineering. L Ron Hubbard did take a class in nuclear physics, but he did not spend his time doing engineering. Scientologists do not spend time studying things like physics or engineering even though the subject is called Scientology. In fact, this one book contains far more real information about physics and this universe than is contained in all of Scientology. Having presented the physics information in this book to the Executive Director of the Church of Scientology, I know first hand that the Church of Scientology considers that physics regarding the physical universe is just something they have no interest in. At least I received a decisive response, which is something I much prefer over no response or a maybe. How an individual can exist in this universe yet claim to have no interest in it is beyond me, but people do have free will and can look at things however they like. After an individual claims to have no interest in the real universe it would be silly to try and affect the universe or complain about it, as do many Scientologists. The Executive Director of the Church of Scientology did answer the letters I sent trying to convince him that he should care about the physical future of the planet and learn a little about physics. Unfortunately, he did not seem to grasp the concepts I presented, though I do believe he tried and that he really was interested even if he could not break free from Scientology tunnel vision.

A Scientologist would make very little forward progress on Scientology levels if he or she insisted that LRH was wrong about an important technical point. So, as far Scientologists are concerned, this is a universe created by individual agreement, not a universe originally created by one God. After people have spent $100,000 and a decade of full time effort learning things like that it is difficult to teach them otherwise. Try and get a committee to agree on anything trivial, much less the creation of an entire universe, and you will soon realize that the likelihood of all mankind agreeing on anything, much less creating an orderly universe, is as close to zero as anything gets. If that is not convincing, ask anyone on this planet to create a new star into this galaxy and you will find that no one can. This is a universe created by one God.

Personally, I welcome any criticism of the knowledge I present. If anyone can create better products or services along the same lines that I am now offering, and will offer in the future, more power to them.

Creating a star is indeed possible. All it requires is a seed sphere of energy at the right location within a galactic cloud. Granted, that invisible "seed" needs to be about the size of a solar system, and it will take billions of years for the actual star to form, but it is possible. It should be obvious that it would take something greater than man, or something greater than a spirit that is interested in a single body, to create something like a star. If L. Ron Hubbard were here today I imagine he would agree with these things and be pleased that someone finally presented a greater understanding of the universe. Since he is not here, and since Scientology is not interested in advancing fundamental understanding, the subject will continue as L. Ron Hubbard left it when he died almost 20 years ago.

Decades ago Einstein proved that energy and matter were equivalent. Why are people still even talking about matter? I mention this because in Scientology lingo this is MEST universe. The term MEST is an abbreviation for Matter, Energy, Space, and Time. There really isn't any matter or time. There is energy moving in space. We use the concept of time to measure that motion, but time is still a concept and not real as is the actual motion. The Scientology MEST acronym is just more slander implying that this universe is "messed" up, and further confirms the lack of formal education in the leadership ranks of Scientology. Today, the chairman of the board is the highest post in Scientology. David Miscavige holds that post. Unless David completed a high school equivalency exam lately, he is still a high school drop out today. I am not being a critic of the choices people like David make in life. How many high school drop outs would turn down an offer to lead a global religion? Listening to David speak, he is a very determined individual who believes completely in his cause. Unfortunately, I cannot reach people like David who do not have at least a minimal understanding of technology. If a person does not even know that the planet has a magnetosphere, how can that person understand the challenges this planet will face much less do something about it. If a person is dedicated to an incorrect notion regarding the relationship between the spirit and creation, how can that person even begin to see truth?

The correct posing of a question often leads to solution, while incorrectly asking a question can bar a solution. The person not asking the right question is not looking in the right direction and lacks the prerequisites to understand the answer. The universe is made of mostly invisible energy plus some condensed energy that we like to call matter. That condensed energy follows the same laws as all the other energy in the universe. In the grand scheme of things the larger energy fields are more important, even looking at solids represented by bound together atoms and molecules. If people really understood Einstein they would have asked where the missing dark energy was in the universe, not "dark matter".

Consider a black hole. The reason that a black hole has so much gravity is that it is so empty. This universe likes to flow into a vacuum. The black hole radiates energy to counter this inflow and so energy is created and the universe expands. There is a singularity at the core of every sun, and a similar gravitational field at the center of every stable sphere of energy. In the case of an atom the gravitational field is not large enough to emit particles as does the sun, even so this concept of a gravitational hole can be used at all size scales. Singularities come in all sizes and commonly overlap, merge, or divide. The idea that two objects cannot occupy the same space is another one of those negative assertions that serve no purpose in science beyond giving researchers tunnel vision.

If Germany and other countries in Europe embrace Scientology they will not prepare physically for future survival in an extreme environment on this planet until it is to late. There are many reasons why Scientologists would not see physical preparation as a solution. Scientologists view the body as a liability rather than as a gift from God. Scientologists believe that this universe is just a mechanical trap without awareness, intended to destroy the spirit. Scientologists have ideas about abandoning this universe entirely in a spiritual sense, and then finding or building a new universe elsewhere. The creator of this universe created the spirit of man. Best of luck to Scientologists who want to create a new universe while denying their own creator and failing to understand how this universe was created. If one does not understand how this universe was created into the void, how could one hope to create an entirely new universe and be a God? From a practical point of view, I doubt that a single Scientologist has even imagined the degree of loneliness and responsibility that the word God represents. Personally, I like this universe. I would not discourage anyone who didn't want to be here from trying to leave. I would advise that people be very sure of themselves before deciding to dedicate their life to that particular path. In this universe of free will each individual is given their own universe of imagination where the laws of this universe do not exist. That universe of imagination is a very real universe running parallel to this universe in the space beyond the event horizon of every singularity. If it is possible for an individual spirit created into this universe to launch oneself into the void, I imagine that individual would be very lonely if nothing else. Could that have been the beginning of this universe? I will leave the answers to the purely philosophic questions open for consideration. Purely philosophic questions are not very practical questions that influence things like the near term future of planet earth.

Maybe this book will enlighten some Scientologists. I do not know of a single Scientologist actively involved in politics, and I have met hundreds. How could a group of people who all think they are God get involved in politics? They would all vote for themselves. Anyway, some Scientologists could end up with exactly what they fear. If indeed life is related to a very high frequency electromagnetic sphere that transcends death, how might a pole shift affect life itself? Could it be that some spirits are pulled into the core of the planet by that electromagnetic whirlpool effect during a reversal? Could it be that some spirits are saved from this fate? That would certainly explain Biblical judgment day.

LRH did not make a single contribution to the advance of physics that I know of. LRH did not release important new products in the fields of energy, rejuvenation, transportation, or any other physical endeavor. LRH focused on spiritual philosophy without first being very sure he new everything there was to know about the practical nature of this universe. If he had focused on completely understanding the universe in front of him first, the spiritual philosophy would have been simpler.

Biblical prophecy talks about false Christs who will lead many people astray. The Bible also mentions anti Christs who will appose Christianity. I have heard that the most advanced confidential Scientology levels contain negative remarks regarding Christianity. I have not read these levels, but I have talked with many Scientologists who have done these levels and each one of these individuals was anything but supportive regarding Christianity. The advanced confidential Scientology levels claim to teach people how to do various miracles like levitating objects or seeing the world through the eyes of the spirit at great distances from the body. I am quite sure I would know if people could do these things, and I neither saw nor heard of any real evidence of those abilities any studying Scientology for several years. I do know that these minor miracles are possible, and I would not be surprised if many people were quite impressed seeing these things.

Judging by my experience, I doubt 1 in 100,000 Scientologists ever achieve abilities remotely resembling what Scientology promises. There is much more to spiritual advance than the founder of Scientology imagined. There are still some Scientology levels that had not been released as of 2003. Maybe some of those levels will enable a few Scientologists to do some kind of parlor tricks that will lead many people to believe that these Scientologists are Christ, God, or whomever they want to be. Christianity and Scientology just look at things differently as far as the big picture goes. In the near term I consider it more important to focus on survival and beginning a better civilization. When that is done there will be time for Jedi Knight training.

What could possibly be wrong with doing some miracles, especially if the results benefit some people? I believe this warning in scripture is more about the purpose or intent behind the actions. Maybe the lesson goes like this. One should not do miracles in a self serving way, trying to impress people and force them into service. Christ taught that it was better to serve than be served. The important thing is raising the spiritual ability of other people so that people can do these things themselves, not to impress people for personal gain. Making things confidential, and charging very large amounts of money that most people cannot afford, does not fit in with the concept of freely given service to the public and honestly sharing knowledge. More likely, all the mystery and high prices are what really attracts people to Scientology who would otherwise see it for what it is if the information were honestly disclosed.

Consider Matthew 17:20. "And Jesus said unto them, because of your unbelief: for verily I say unto you, if ye have faith as a grain of mustard seed, ye shall say unto this mountain, remove hence to yonder place; and it shall remove, and nothing shall be impossible unto you."

The disciples were sent out to do some good works, making insane people sane, etc. They came back with excuses, saying they couldn't do these things, implying that they were not as capable as Christ. Jesus made it quite clear in Matthew 17:20 that the disciples were indeed capable of these things, and even greater miracles than Jesus had done. Jesus did not levitate mountains.

Maybe it is tough to have the purity of faith present in a mustard seed. Faith implies that God is part of the picture. It might require the participation of God to move a mountain purely by thought projection. There are many ways to move mountains. Certainly, a reversal event will see major and rapid mountain moving. Using advanced technology, one could cut away blocks of a mountain and move it piece by piece. Super advanced technology could probably cut away the base and pick up the entire mountain. That doesn't rule out moving the mountain simply by thought projection. The important point is that Christ taught that we all have at least as much or more potential as him. Some Christians may not understand that, but it is true, that potential just has to be realized.

Revelation is full of predictions that appear likely in the context of rapid environmental change. A great battle is predicted called Armageddon. In desperation, countries could decide that war is their only option faced with a shortage of resources. Even without extreme environmental challenge the political situation around the world today is not very stable. Considering the distribution of military forces and allies around the planet, it wouldn't be much of an Armageddon if that battle involved the U.S. and allies versus Islamic counties in the Middle East. The way the Bible talks about Armageddon it would seem that for an event to match that prophecy it would have to be a more significant battle. The weaponry deployed in the last brief war against Iraq did not represent the true capabilities of the U.S. military. To some extent, the capabilities on the cutting edge of U.S. defense are being kept confidential. The most advanced weaponry really was not needed against Iraq. Considering an Armageddon scenario, even if China joined with the countries in the Middle East the outcome would not be much different. That is extremely unlikely though, since China is growing closer to the U.S. as time goes on not farther away. In my opinion, China knows what the U.S. is capable of and for that reason alone would not attack the U.S. even if driven by desperate environmental conditions. So, in the near future anyway, looking at how things stand now, the only Armageddon scenario that could challenge the U.S. would have to be an extraterrestrial threat. I am not predicting that, I am just saying that if scripture unfolds as accurately as it has many times in the past then such a threat should be considered.

I do not expect Europe to begin taking action until a catastrophic environmental threat is all to clear. In the news recently, France moved many of their art treasures to higher ground, anticipating a 100-year flood. France takes this action after seeing the flooding in Germany. The last 100-year flood in France was during 1910, so France believes they are over due and has taken a precaution.

France could go years worrying about a 100-year flood that might or might not hit. Then, France could suddenly experience an 800,000 year flood, combined with earthquakes and other natural disasters. The process of preparing for an 800,000 year flood, versus preparing for a 100-year flood, is completely different.

I expect a small number of Europeans will realize what the future holds and want to move to the U.S.

Eventually, the majority in Europe will realize what they face. At some point the increase in earthquakes and volcanic activity will be too great to ignore. Having waited to long to do anything effective, a sense of desperation will set in. At that point, strict controls could be placed on the material resources that could be used to build survival systems and products. Some kind of system could be established to physically and permanently mark certain people, authorizing those individuals to buy and sell certain things. The "Mark of the beast" could actually come to pass. Maybe this future can be avoided. Judging by the track record of Biblical prophecy, I wouldn't bet against prophecy.

We cannot know the exact details of the future, or the exact timing of events. The best we can do is prepare for the worst case, and hope that we are better prepared than we need to be. The challenges are defined. It is best to move rapidly towards a solution. Once a survival solution is in place there will be the luxury of time to focus on other things.

If free will is so important, why do we have prophecy? Is the life of an individual driven by destiny or free will? Destiny implies an ordained future, which does appear to be at odds with free will. The idea of destiny is that forces are at work that are greater than the individual and drive at least some events of daily life. Destiny would not be destiny if it were not difficult to avoid. Personally, I have tried putting that to the test. Destiny can indeed be difficult if not impossible to avoid. I believe that some individuals sacrifice a certain amount of free will voluntarily so that others may have greater free will to pursue personal dreams.

Both free will and destiny combine to shape the future. From a personal point of view it is best to just focus on free will and personal planning to achieve particular goals. One does not go looking for destiny; destiny will find you as need be. Personal accomplishment contributes to the role one plays in destiny. I liken it to an agent sent out on a mission briefed on a need to know basis. Say that agent was a great pinball player. The secret agent is sent out on the mission. Part of the mission is to enter a pinball tournament in Mexico and impress a lovely lady by winning the tournament. She would be the agent's next contact for the second phase of the mission. The agent isn't particularly concerned about the tournament, and plays a calm game winning easily. Later, the agent discovers that the fate of the world hung in the balance, decided by the outcome of that pinball tournament. Say that the agent's handler new that this agent did not play pinball very well under pressure but that he loved to impress girls with his game. Should that agent have been told the whole truth upfront?

Two thousand year old prophecy, having been through many translations, cannot be precise when it is read 2000 years latter in a different language. That leaves prophecy like the book of Revelation and some of the New Testament open to interpretation. Not only translations, but also the limitations of the original language limit prophecy meant to describe real events in a world very different than it was 2000 years ago.

There is a popular movie that was released called "Left Behind". Visiting Barnes and Noble I noticed that some books in the "Left Behind" series sold as many as 30 million copies. Correct or not, the Biblical interpretation promoted by the "Left Behind" series is popular.

"Left Behind" begins with a bunch of people who disappear without explanation. The people left behind compare notes. They decide that the people who disappeared did things like attend church regularly, and prayed often. I have only watched the movie; I didn't read the whole series. In any case, I did not get the impression that the elect took any real action, or did any prior planning or preparation to endure a temporary extreme environment. The movie made it look the elect just disappeared into the ether. The "Left Behind" movie promotes a particular Christian denominational belief regarding salvation.

The Roman Catholic Church believes in salvation through good works. Some denominations prefer to consider action secondary, assuming that God asks nothing practical of "mere mortals". The beam me into the ether concept fits the philosophy of protestant denominations. That doesn't mean that interpretation of the Bible is correct. Personally, I am still a Roman Catholic, and I believe in salvation through good works.

The life of Christ was a life of action. Christ said that we would know people by their actions. The idea of coming to the father through Christ is about emulation to some degree. What else could it mean? Emulating Christ means taking action. In the days of Christ more than one Pharisee who attended church regularly and prayed in abundance helped put Christ to death. Granted, there was purpose in the death of Christ, as much as in the life of Christ. That doesn't absolve the Pharisees. A verbal statement of faith combined with regular church attendance isn't much different than what the Pharisees were doing before they put Christ to death. Would it be much different if people failed to take action for a new civilization, allowing their neighbors and themselves to be destroyed, possibly even losing all of mankind, all because of an incorrect interpretation of scripture? A verbal profession of faith is not sufficient. Consistent action is necessary to live a life of joy and forgiveness in the eyes of God.

Christ asked that people make preparations for his return. Considering Christ went into great detail concerning the nature of things during the time of his return, can there be any doubt concerning the nature of the preparations that would be required? Putting ones house in order could be taken very literally. However it is interpreted, the practical nature is obvious. The language of that day was capable of differentiation between the temporal and the spiritual.

From a logical point of view, it is better to go with salvation through good works as an interpretation. If by chance good works are not necessary, it doesn't hurt to do them anyway and prepare for a future of life. Being wrong on the flip side, doing nothing and then suddenly realizing that something was required, one could regret that interpretation of scripture.

I mention this because there are millions of Christians in the U.S. who have decided that they know best, and that obviously a beam me up Scotty rapture is the way things are going to work and so only "crazy" people like the Branch Davidians would try to make any physical preparation. I am not saying the beam me Scotty idea is impossible. Certainly it is possible. At the same time, I would not assume that is what is going to happen, and even if it did it would have to be limited to available resources if that rescue is going to physical. Revelation talks about God's place being with man, here on earth, building a new civilization. If the idea was to leave the planet I am sure something would have been said directly related to leaving the planet in the Bible.

Assuming that one knows every detail of the mind of God is a risky assumption at best. Christ did not presume to know every detail, nor the exact timing of God's plan. I suppose that "Left Behind" is an ongoing series so the conclusion of the series could present a different explanation for how the people disappeared.

Two words in Revelation, "taken up", do not necessarily translate into vaporization and merging with the ether, or a beam me up to heaven scenario. There are other possibilities. People in trouble could be air lifted from certain areas and set back down at a safer location. Even this technology would be magic to a Christian 2000 years ago.

All of these interpretations focus on the physical body, which is why I believe that they are all incorrect interpretations of the rapture. I can think of no greater rapture than achieving oneness with the mind of God and knowledge of all creation. A recognition of personal spiritual existence, and the fact that the body is nothing more than a temporary tool easily replaced if necessary, is probably closer to what the original Hebrew text of the Bible was trying to convey. That kind of spiritual rapture can happen at any time, is not limited by available rescue craft, and has nothing to do with the body whatsoever.

The idea that one has to physically die to go to heaven, hell, or anywhere in between is silly. It is like comparing a slice of cheese pizza to different planets. The two are just not that closely linked. If one has not achieved spiritual freedom in life, then the death of the body is not going to accomplish more than revealing the true state of the spirit, as it was in life, but now without a body in the way. The spirit always exists along side the body with the potential of being one with God and experiencing all creation at will. Heavenly rapture can be achieved in life while one is still controlling a body, or after death when one is not controlling a body. The significance of physical death has been exaggerated. It is more significant when the entirety of mankind if faced with possible destruction, and at least some people need to be physically saved so that mankind continues to exist.

Even the spirit of an animal continues on after death, with the one option of being reincarnated to build a body of that same species. Just as the spirit of a planet is different than the spirit of a bear, so is the spirit of man different from the spirit of an animal. The spirit of man is not focused on building a body up cell by cell. The spirit of man was created with greater potentials, like being aware of God, understanding creation, and acting as a steward of creation.

An individual spirit could be banished to the core of a planet or sun if that individual spirit was not capable of truly conceiving of God. Knowledge and truth are indeed power. A true spiritual recognition of God, and a spiritual knowledge of creation, is freedom from the traps of this universe. A failure to recognize God is like someone living in a bowling alley who has never been outdoors and fails to recognize the roof overhead. That poor spirit would have little chance surviving outdoors in a hurricane. In fact, that poor spirit would not even be able to conceive of a hurricane and would probably laugh if someone tried to warn them about it much less tell talk to them about God.

Here is quote from Revelation 3:16, "knowest not that thou art wretched, and miserable, and poor, and blind, and naked." There are two possible interpretation of naked. The first is physical and the second is spiritual. Does it makes sense that a bunch of people would walk around physically naked and not know that they are naked? Maybe prior to the Garden of Eden, but not during modern times. That leaves only the spiritual interpretation. There are many ways to describe this from a spiritual perspective. The best way to visualize it knowing what people know today is to describe the spiritual existence of people today as a weak naked singularity that could end up being trapped in the core of planet Earth for the duration of this universe during an electromagnetic reversal cycle.

Not everyone will be able to understand the concept I have just presented, and the people who do understand will probably not end up being trapped in a planet. Even so, I will present some advice. One way out is the way in. The core of a planet or sun is a little gravitational black hole, certainly not naked because a planet surrounds it. Nothing enters a black hole. Any energy field that manages to approach close enough ends up merging with the outer magnetosphere and pervading the entire planet or sun. I say manages to approach close enough because there is a great deal of energy radiating from the center driven by fusion events and neutron production at the core. The best course of action would be to recognize that God pervades this entire universe, and then become one with God for a little while and then emerge spiritually at any point in this universe. That could be called teleporting. Atoms sometime experience quantum tunneling, disappearing from one point in space and reappearing at another location usually close by. Quantum mechanics also recognizes particles that appear out of nowhere in space for brief periods of time, this are called vacuum events. The concept is similar looking at an individual spirit. This entire universe is life; small life, large life, and life of all different kinds and sizes in between.

I doubt that anyone who understands these things will ever end up trapped in a planet. This advice is something that an atheist could not imagine, so it would seem of little use mentioning it. Maybe it will help someone who reads this book and decides to sit on the fence and wait on taking sides until it is most difficult to do so. It would be most difficult for someone suddenly realizing that they are a spirit, in need of some help, after failing to accept that help when it was possible to provide it. Only God could reach into a planet or sun and extricate an individual spirit if that spirit is not able to do so.

The reason that people fall into the traps of this universe, and the problems of life in general, is an unwillingness to freely experience what this universe has to offer. Placing a label of "good" or "bad" on a particular personal experience, then spending all of ones time chasing and resisting certain things, is a problem. There is nothing wrong with chasing or resisting itself, the problem is not being able to simply be at peace and experience anything just as it is without chasing or resisting. The problem is that one can become trapped in any situation if one is not able to step back now and then and simply see things for exactly what they are without throwing in ones two cents all the time. This is true for someone stuck in some situation in life, just as it will be true for some individuals on this planet who find themselves spiritually trapped for eternity. The individual that is unable to just experience incoming energy, and know the universe for what it is, instead resists incoming energy by emitting counter energy. That very act of constantly using energy ties a spirit to a particular location and makes that concept of teleporting impossible to achieve.

Some Scientologist are aware of the spiritual nature of individuals. The subject of Scientology misleads people, and claims to have a solution resulting in complete spiritual freedom. That claim is false. The subject of Scientology fails to accurately describe this universe, and fails to present a practical spiritual solution to the challenges of the future. People have already spent decades following the path of L. Ron Hubbard, and those individuals aren't much closer to true spiritual freedom than when they started. The individual spirit on this planet needs a solution in the time frame of years. Faith in God is one solution. Really understanding creation would certainly help.

The moral relativism of Scientology, implying that God is whatever a particular individual happens to decide God is, is not the solution to spiritual freedom.

God exists, and there is one God for this universe. A Scientologist claiming otherwise, or claiming to be God, is little more than a snowflake crying out in the middle of a solar flare on the surface of the sun. The Scientologist confuses the universe that we all live in with the freedom each individual is granted to imagine their own universe in their own mind. On top of that, L. Ron Hubbard further confuses Scientologists by claiming that the real universe was created by the agreement of a bunch of separate individuals. The most misleading false information always contains a grain of truth. In the beginning, there was only God, and God created this universe into the void. God also created the individual spirit giving each spirit a little piece of the power of creation that is God. Prior to the separation of the individual spirit from the mind of God, one could argue that the individual was in "agreement" and helped create the real universe. That is a poor argument though, because the statement of the argument implies that the individual did not exist as an individual at that time. God still exists, and is far greater than the combined strength of all individuals separated from God, including planets, suns, and galaxies. Denying that existence is like spitting into the wind of a thousand hurricanes. Scientologists who fail to see this truth now will have an eternity of existence in pain latter to reconsider. At that point, the individual spirit will have only two options; continue an existence of pain or surrender individuality and become one with God again. An individual spirit who has lost faith in God, and who has further lost faith in themselves having become trapped, will have great difficult finding spiritual freedom in God after falling to such a low level.

Free will is a two edged sword without which this universe would be a very boring place. People are welcome to call themselves God, damn this entire universe, and say they could build a much better universe any day of the week. If those same individuals find themselves damned, and unable to make good on their boasts, who will mourn for them? This entire universe is a universe of evolving life. Events like planetary, solar, of even galactic magnetosphere reversals are as much spiritual events as they are physical events.

The previous paragraphs represent a spiritual interpretation of rapture versus damnation, or heaven versus hell. I have heard many interpretations of the Bible that are very superficial and focus on a physical interpretation whenever possible rather than a spiritual interpretation. Sometimes people call these physical interpretations "litteral" interpretations. Language did not exist to write a detailed spiritual version of scripture 2000 years ago. If someone insists on a physical interpretation of the rapture that interpretation should focus on known physical technology. Even the body has a physical and a spiritual component. The physical body is not going to teleport into the ether without physical technology capable of teleporting bodies through the ether. Mankind needs to go through a step by step process to acquire advanced technology. Technology cannot truly be owned without understanding all the building blocks that lead up to a major advance. From a physical point of view, the elect may be those people who find themselves in a difficult position. Some people who are better prepared for difficult times may end up doing the electing. How would you feel if you were the warrant officer sent to rescue 100 people threatened by a lava flow with a helicopter that could not lift off with more than 12 people? What if those people on the ground had 30 minutes at most to live, and the nearest safe location where you could drop people was 20 minutes away flying at a speed greater than the manufactures rating for your aircraft? That scenario is what a physical interpretation of the rapture would involve.

I prefer a spiritual interpretation of the rapture that involves spiritual transcendence of the body, and being embraced by the grace of God that fills the sky and everything else in this universe. That kind of rapture is unlimited and personal. That could happen for anyone at anytime, it does not require physical death, nor is it limited to how many bodies can be packed like sardines into a limited number of aircraft or spacecraft. Spiritual rapture represents an understanding of God, and the ability to become one with God while still retaining individuality.

Physical salvation should be the domain of group efforts and personal labor, as it was with Noah. The problem is that many people have confused the physical and the spiritual. Some people believe that they need not do anything practical to save the body, and that they have to do all kinds of practical things like attending church and making donations to save the spirit. The reverse of that is actually the truth. Saving the physical body requires work just as Noah was asked to do some work. Saving the spirit requires understanding, faith, and spiritual transcendence. These things of the spirit have nothing to do with the body, or anything else that can be seen, touched, heard, or spoken in this universe.

If an individual does not care about the body, then I suppose that individual could do nothing in terms of physical preparation. That would still be a mistake though. Suicide against the physical body is not as bad as denying God, but is still negative. If an individual gives up on physical life without good reason would it make sense for God to entrust the management of any aspect of creation to that individual? I don't think so. Suicide has many faces. Sometimes people go down self destructive paths, failing to see the warning signs, and failing to take effective action to preserve their own life and the lives of others.

Here is common sense look at one popular physical interpretation of the rapture. The litteral interpretation is that some people who happen to be in fields will climb up some ladder of be beamed up to some space ship while the person standing next to them in the field will be left behind. Of course all the people who happen to be in cities at the moment the rapture takes place are completely out of luck unless they can find a nearby field and run there fast enough. I am being a little sarcastic just to make it clear that it is virtually impossible to defend completely literal interpretations of anything. It just is not possible to understand something being completely literal. Look in the dictionary and you will see many definitions of each word used in any sentences. If nothing else, understanding is required to select the definition that will be used to understand any sentence. The absolute minimum thing that any responsible reader should do is to ask whether or not the scenario presented by a particular written communication makes sense. For example, if people needed to be physically rescued from a looming physical catastrophe, what would happen if someone were to fly around trying to pick certain individuals out of the fields and leaving the other person standing beside to die? I guess that would trigger quite a few physical fights to the death to climb aboard some rescue craft. That would be counter productive if the idea was to physically rapture people. If no fight to the death happened, then the people willing to be left behind and physically sacrifice themselves so that another may be rescued could be considered the nobler of the two. A purely physical interpretation of this scripture does not make much sense, therefore a spiritual interpretation should be considered. The scripture just says that the two people were standing in the field and one was taken up. The scripture does not say that after the rapture only one body is standing in the field where there were two. People assume this even though it is not written because they cannot conceive of the spirit separate from the body. Now, before people further misinterpret what I have just written, do not assume that a spirit taken up means that the body must die. That is another assumption, and a false assumption at that. It is actually much better to control a body at a great distance via spiritual links through space. The individual spirit that remains trapped in a body is severely limited by the amount of power the individual spirit can wield since to much power to close will destroy the body.

Prophecy is given to mankind in visions. Those visions have to be described in words, which is not always easy to do if part of the vision is something invisible like a spirit. After the vision, the prophecy has to be written down and then translated through a number of languages over time to arrive in the hands of people today. That is why a little bit of common sense and judgment need to be kept in mind reading scripture, especially prophecy. Even the English was very different 1000 years ago. Given a high enough level of understanding there is no need for prophecy.

What have we done so far as individuals, and as a nation, to physically prepare for the events described in Revelation? Mankind has made some progress. Two thousand years ago technology did not exist which could have helped people survive through a pole shift. I believe Noah faced a magnetic excursion, which is a much less challenging event. Building an Ark like Noah did is not sufficient to survive a complete magnetic reversal. Forty days and forty nights are not that long to endure, especially if there isn't a problem with volcanic clouds of ash and cosmic radiation.

As a nation we have advanced technology at a torrid pace over the past 200 years. We do have the technology to endure the kind of environment I describe, even in relative comfort if people prepare well. We have the capability of storing seeds, preserving livestock, lions, tigers, bears, etc, as Noah did. In addition to technology we have a free democratic nation with an economic engine unmatched across the globe. The U.S. is capable of preparing for very difficult times and emerging from those dark days into a bright new world with a civilization that everyone can be proud of. Having the potential to accomplish something, and actually accomplishing it are two different things. The purpose of this entire book is to motivate people towards survival. I am confident that a small group and myself will fare well, even if I have to personally take care of everything for that small group. It would best if other people took positive steps in the right direction and encouraged some level of public political action. Much more can be accomplished by any one individual in the context of national support as apposed to trying to do something in the context of a society that is resisting forward progress.

Economic expansion is driven by a collection of individual attitudes. During World War II the U.S. rose to the challenge, economic output skyrocketed, and the country emerged victorious. That attitude needs to return, even though the challenge is not as obvious as German U-boats attacking private shipping fleets.

It is more important what people do, not so much that people believe or disbelieve the future as I say it will be. Technology like Sky Cars need to be deployed rapidly. If people in congress vote for a new air traffic control system based on a traffic argument rather than a global environmental change argument, that is fine, whatever it takes to motivate people to accomplish certain projects.

Here are some more predictions for the future.

The U.S. economy will boom creating wonderful new vehicles, survival systems, rejuvenation centers, and self-sufficient local energy units. The sooner these things begin the better, enabling more people to prepare a self-sufficient life style and achieve real economic freedom. This boom will be temporary, lasting until the environment becomes so severe that it takes a toll on economic production despite the efforts of hard working imaginative people who will do their best in the face of any challenge.

Americans will join together with friends and neighbors across the nation, creating the greatest political movement this nation has ever seen, and electing individuals from among those friends and neighbors who will lead this nation into the future. Well-funded minority groups that are not driven by an honest respect for the well being of the general public will lose political control. Certainly some well-funded groups do care. The real change will be that the public will learn to tell the difference between a fronting façade and what is truly in the hearts of individuals and groups. People will view actions in the light of a higher understanding, rather than being swayed by rhetoric. In brief, the majority of Americans will be less confused, more goal oriented, and not easily fooled by throwing a bunch of money at TV spot advertising. An open forum debate or talk show on TV can help reveal intentions, since responses must be candid. Still, people will watch less TV.

The days of dependence on foreign oil and non-renewable dirty energy sources will end, to be replaced by clean renewable energy in the hands of individuals and families.

The days of congested freeways spewing toxic fumes will end, replaced by clean air transportation delivering people safely and rapidly to their destination without need of roads or special airports. Roads and rail may still be used for commercial shipping, but safe personal air transportation will grow to dominate the future.

Many Christian denominations will come together, inspired by projects put forward by the ecumenical movement. The word Catholic means universal. Vatican III will unite Christian Churches around the world by defining a set of mutually acceptable compromises. Many separate Christian churches will consider themselves part of the Universal Church. That does not mean that the organizational structures of these different churches has to be merged. What is important is that these different churches put aside petty difference and work together towards common goals. I will not give precise times on any predictions I make because the timing is always influenced by delays amounting to the exercise of free will. Particular calendar dates are far less important than actual events and the sequence of those events.

Getting into a deep understanding of prophecy, one realizes that prophecy is true because it is made to be so, or known to be so. That includes prophecy that spans thousands of years. Certainly the book of Revelation was written through inspiration from the Holy Spirit. In other words, God works through the spirit on occasion. The Holy Spirit is better understood as a plural, "the Holy Spirits". Angels, saints, and even the converse are all very real. God can work through any spirit in this universe. If a prophecy involves a physical object, you can bet that a physical object will have to be made by the hands of someone to fulfill that prophecy. For example, if we interpret Revelation as saying that a city floating in the air is going to exist on earth then someone is going to have to build a floating city. That is certainly technically possible. The amount of effort and technology is at least many decades, if not hundreds of years away, looking at where mankind is now. So, to make that prophecy a reality, either very many decades would have to pass, or some friendly group in the galaxy with a pre fabricated floating city would have to deliver one. The Bible doesn't say exactly how a prophecy will come to be, or exactly when, only that certain events will happen.

As more people come closer to God inspiration will flow into the minds of many people advancing culture and technology in ways that can only be dreamed about now.

Evil and ignorance will dwindle as more and more people begin to see the truth of things in front of themselves clearly. Having the knowledge to set things right, people will do so, working with political groups, social groups, or as individuals depending on the situation.

With the restoration of the protective mode of Earth's magnetosphere, health and vitality will increase dramatically across the planet, not just for mankind but also for all life. Earth will prosper and become a hub of galactic trade. Life will become a thing of joy, challenge, and excitement for people as it should be for people today. Many people today consider that the challenges life presents, just getting by on a day to day basis, are to great in themselves so people lose hope and enthusiasm for a better life. That attitude is what creates the whole situation of day-to-day life being difficult. The reason things are so difficult is that virtually no one is doing anything to change the bigger picture and make things better all around. That is the only way to make life better at the personal level. Until a majority of people change their attitude, and decide they are going to do whatever they can to make society better no matter how difficult it is at first, things will just continue to be problematic for most people.

Consider another prophecy from the book of Revelation. A movie was released called "The Gangs Of New York" set in the early 1800's. The characters of that movie are anything but lukewarm. It may be that none of them were particularly right. Even so, every character had a certain faith in God, and a willingness to take action based on principle rather than living in fear or complacency.

Here is Revelation 3:16-17, from the King James Version of the Bible:

"So then because thou art lukewarm, and neither cold nor hot, I will spue thee out of my mouth. Because thou sayest, I am rich, and increased with goods, and have need of nothing; and knowest not that thou art wretched, and miserable, and poor, and blind, and naked." Can you see through the eyes of the prophet visualizing today's society of somewhat arrogant people believing they have conquered the universe only to be nearly annihilated a short number of years later?

I heard someone asking about the meaning of this scripture on a Christian broadcast station. The caller asked if the ecumenical movement was lukewarm. The commentator said the ecumenical movement was not lukewarm. The question itself demonstrates great confusion concerning this scripture. There is a little symbolism here, particularly talking about God. The engines of creation, or magnetospheres throughout this universe, could be compared to the mouth of God in the absence of any better terminology. Looking at the second verse, I consider being lukewarm an attitude that people take assuming they need no help from God. As a society progresses, with modern farming techniques and a little factory production, people start considering they need no help and are more than capable of handling this entire universe all on their own. It is that arrogance and ignorance, failing to realize that one does not know, that is the greatest threat to mankind.

Some day mankind may build a civilization that could enjoy the fire works of a planetary pole shift, selling the best suites in floating cities over looking the hottest new volcanoes. Today, mankind can barely put a craft in earth orbit much less travel to the stars and back in a matter of earth days. Today, mankind lives a short life, aging and in pain for more than half of that short life. Until mankind can prove otherwise, a wise man would have some humility measuring the capabilities of mankind in the 21st century rather than over confidence.

Revelation was written with a very specific purpose in mind, directed at a very specific time in history, and very specific people. That time in history is now. Those people are you and I along with everyone else who will be needed to change things for the better. How many people today are increased with goods, and consider that they need nothing from God in the goods department? This passage is talking about additional practical goods. Most everything people own today will be of little or no use in an extreme environment. All the stuff we own which could have limited use runs on non-renewable centrally distributed energy. This passage really can be taken quite literally. Look around you today. How many people can you find who are just not feeling very well physically or spiritually? How many own products or clothing that will be of use in an extreme environment? Finally, how many are blind to the truth of God, the spirit, and all of creation, unable to directly see the most important spiritual nature of creation, which is in fact more than 90 percent of this universe? Granted, mankind has technology to measure and visualize a planetary sized magnetosphere, but how about understanding what is being measured? That understanding is true insight.

God does have a vested interest in mankind. Despite all the debate about evolution, the body of man and women did not suddenly and instantly appear. We see evolution in the geologic record. There is no reason why that record would lie. Mankind is highly evolved to use technology for survival, and thus take on the role of stewards of creation. It would be a set back for the creator of this universe if mankind was annihilated. That does not mean that God is going to change the whole design of the universe to prevent energy releases during reversals of a magnetosphere. It does mean that some effort would be made to rapidly advance technology, and that prophecy would be written to try and warn people even if they could not understand the situation at the time the prophecy was written.

What if Noah had a different attitude? "No thanks, I really don't think I need a big Ark in the middle of this desert. I think I'll go back and live comfortably in my house. I am sure Ark building would be very difficult work anyway, and a complete waste in this desert. I am sure people would call me crazy. Why did I hike out into this desert anyway!"

Many people will never be asked to put their own life on the line for a neighbor. How about publicly supporting the creation of a better world? The information in this book is important. A great many people can be helped through the difficult times ahead, but only if each person takes some action to prepare for the future. It really is a win all lose all scenario. One person not actively participating is one person closer to not enough people to take public action. Mankind will survive into the future, even if it is just a small group going forward to repopulate the planet. One cannot expect any public support if one is not willing to support a campaign platform that is geared towards creating some kind of public programs. Those programs simply do not exist today, and projects are not going to just magically appear without overwhelming support to create them.

The generalities of the future are in the hands of God. The details are in our hands, modified to a greater of lesser extent by faith and destiny.

Throughout this book I have been careful to present interpretations of scripture as my own beliefs, opinion, preferred interpretations, etc. Very good reasons support my interpretations, but I still present these things as my own interpretations, as should anyone who tries to interpret the word of God. To many people have launched campaigns of destruction like the inquisition and "ethnic cleansings" claiming to be doing so in the name of God. Thessalonians 5:9 says "For God hath not appointed us to wrath, but to obtain salvation by our Lord Jesus Christ." Some people ignored that part of the Bible and many other scriptures that are simply stated when starting in on jihads or persecutions. The wrath that is spoken of in Revelation is the wrath of nature for the most part, though some people who fail to prepare for the future can be expected to bring about their own destruction. Mankind may have influenced the outpouring of this wrath via nuclear testing, oil drilling, and generally living in disharmony with nature. Obviously, people in leadership positions have misinterpreted scripture in the past, is there good reason to believe that this is not still happening today? That is why it is best to allow people to arrive at their conclusions rather than try to force a certain belief.

A temptation will divide people into two camps in the years to come. Releasing technology like rejuvenation and low cost energy gives people a choice. People could use their additional free time and regained youth to lapse into sloth and spend most of the day chasing skirts, or being chased as the case may be. Or, people could work harder now, knowing that it is important to prepare for the immediate future and repair the mistakes of the past. That is the smart thing to do recognizing the an eternity awaits for long vacations, and also realizing that eternity could be lost by failing to prepare for a near term transition today. My greatest hope is that these days of challenge will be shortened. Three or four years to endure would be a blessing, considering that there is no reason to rule out the magnetosphere staying in a circular current mode for 300–400 years. These events have been very unpredictable in the geologic past.

In addition to common sense and judgment, it is a good idea to look at everything related to a particular prophecy at the same time. How about the return of Christ? Does arriving on the clouds mean that changes in the sky will herald an arrival, or will the actual arrival be a dropping down from the clouds? Is this arrival on the clouds a spiritual arrival or a physical arrival? Is there any truth to the prophecy that Christ will return like a thief in the night, implying an unexpected timing or unexpected method of arrival? Does the return of Christ even have anything to do with one particular individual? When I hear people say they are looking for someone to drop down through the clouds with "CHRIST" written in big letters all over them I wonder if these people have not missed the point. Someone claiming to be Christ descending in military spacecraft, capable of taking control of the planet by force, wouldn't be much different. It is an interesting exercise asking people exactly what they are visualizing when they profess to believe a particular prophecy.

One person does not make a civilization. It is more likely that the return of Christ is about the hearts and minds of a group of people changing, and becoming more like Christ as an ideal, or in a spiritual sense. A long list of seekers, both past and present, contributed to the answers that I present in this book. That list is long just looking at major figures in the development of physics, exploration of the world, etc, etc. The list gets longer adding all the "ordinary" people, husbands, wives, friends, family, and social groups that provided the environment making it possible for all those seekers of new knowledge to find what they were looking for. How about our armed forces that have given us all a free country past and present to live in? In the eyes of God I am sure that this bigger picture is far more important than the physical life of any one individual. It is also clear that many people over many generations contribute to the advance of technology and knowledge required to understand creation.

If some extraterrestrial in human form descended from the clouds, either like superman or in some high tech spacecraft, claiming to be Christ, would that satisfy people? What if we were asked to surrender things like democracy, personal freedoms, or national sovereignty? That would make me wonder if some alien civilization wasn't just using their own version, or interpretation of scripture, to encourage the whole planet to surrender without a fight. I personally believe in democracy and freedom. I love this country and the very ground it is built on. I would not abandon it and fly off in some alien ship even if the environment becomes difficult for a little while. If star ships exist, there is no reason why the United States of American can't learn how to build star ships.

Deep in everyone's heart people recognize what is truth, and what is doing right by other people. People know that it would be foolish to sit back and do nothing in the face of looming catastrophe. At the beginning of this book I started by asking a question. What if you knew the reasons behind Biblical prophecy? What if your friends had all the best intentions, and professed a belief in scripture, yet they really did not understand it and that lack of understanding caused them to wait and see if anything would really happen rather than being proactive. Knowing the magnitude of the challenges people would face, and knowing that even with complete participation and proactive action many millions would still lose their physical life at least, can you see why a week of tears would stream down my face? Surveying people who read this book, most people take it all rather lightly as if looking forward to a fireworks display if anything. That is fine. Action is much more productive than grief, and the right preventative action always makes one feel better. The thing that bothered me most at the time was that only 1 other person really understood the situation, and he considered it hopeless to try and do anything to help people, leaving me to try, and I was not in a position to do anything quickly or on a broad scale at the time.

If you cannot see these things, or still remain in doubt, then perhaps ignorance truly is bliss. At times I almost envy that ignorance, though I do not envy the price that will be paid for that near term bliss. If you do understand these things, and decide to help, I leave it your hands to teach others as quickly as you can. This book is available from many retail outlets, and I have made the contents of the entire book available as HTML pages on the book web site. In publishing this book I have done everything I can as far as reaching people goes. I plan to focus on the development of new energy products for houses and vehicles next. Once that is done, and I am confident that will not take to long, I will then focus on a political campaign if it looks like enough support exists to do so.

Only history will reveal the details behind events described in prophecy. Maybe a physical rapture by UFO's and an extraterrestrial born Christ, building a better civilization on earth after everyone left behind is destroyed, is the way things will go. I do not believe that interpretation, though I understand how some people could get that impression. I believe that there is room in the arms of God for everyone, not just how many people can fit in a space ship. I also believe that with some effort preparations can be made for a majority to endure difficult times on this planet, and furthermore that those preparations should be made rather than assuming that nothing should be done and relying on a fleet of alien space craft will descend and pick people up.

Here is another common sense interpretation of scripture regarding the return of Christ and salvation. First, clearly, failing to recognize the reality of the return of Christ and instead embracing any number of false interpretations or false individual claims will result in a failure to achieve salvation. Reading scripture, that recognition of truth is not going to be obvious. Many people will claim to be Christ, or God, and it might not even be that the return of Christ described in scripture is about a single individual.

The Bible says that Christ is the way, the truth, and the life. Not one of these qualifications requires that Christ be a single individual. What if salvation could be compared to finding a particular small gold nugget buried deep in the ground some place in South America? Finding that nugget quickly would be impossible without a map. Truth is what remains after a virtually unlimited quantity of false information is stripped away. A great many different churches today preach a great many different versions of salvation. They cannot all be true. What is a "way" if not a path to be traveled? Coming to the father through Christ implies some kind of action. How can one go "through" without walking some path? The Bible is not talking about a physical path. Walking from a point labeled "A" to a point labeled "B" and then mouthing a few words at point "B" is probably not the path to salvation. The only interpretation that makes sense is traveling a spiritual path separate from all the falsehoods of this world, and seeking greater personal knowledge combined with an understanding of God. Only God can judge whether or not an individual spirit has traveled that path. Only the individual can know if he or she is trying. I believe it would be a great mistake for anyone to assume they have arrived at the end of that path to salvation. That would be like assuming one is God and deciding to favorably judge themselves.

One more prophecy interpretation before looking at extraterrestrials and UFO's. There is a prophecy about the dead rising to life. Does that mean moldy skeletons are going to be animated and claw their way out of the ground? If that were the case, the people who received the vision of that prophecy would have tried to be more specific describing skeletons clawing their way out of the grave. The prophet would have put some emphasis on the word "bones", or described flesh being attached to the old bones from the grave. As silly as this sounds, I would not be surprised to find that this is the interpretation most people would give when really pressed for a detailed interpretation of this dead rising to life prophecy. Surveying a couple people, I found that this is what they believed. Prophets do try and convey prophecy as best they can from the visions given to them. Today "modern" medicine extends life in a very unnatural way. All the drugs and artificial this and that do little to improve the quality of life, beyond keeping someone breathing who would have died much earlier in a more natural environment 2000 years ago. Seeing the physical condition of some elderly people today in a vision, and seeing a state of physical decay beyond that of anyone the prophet had seen put in the grave during his day, and then seeing some of those individuals rejuvenated, the prophet might get the impression that the dead had risen to life in a physical sense. Just seeing a vision of some elderly people rise from bed one day and walk around, a prophet of 2000 years ago might have written that the dead had risen to life. Seeing millions of such individuals rejuvenated in a vision, the prophet might consider that the dead of both the present and all the past had risen to life. If a prophet believed in reincarnation that would be true in a way, and would have influenced how the Bible was written. People who do not believe in reincarnation, and who do not understand the nature of the spirit in the same way as the people who wrote the Bible, would interpret scripture much differently than was intended. My point is that there is more than one way to look at Biblical prophecy, and that no one should try and force a particular interpretation on anyone.

The Bible says that we are all created in the image of God, so every individual should be able to come to their own conclusions regarding prophecy without relying on a second hand interpretation that is a far cry from the understanding of the prophet himself. When it gets down to it, without having the prophet at hand to question in detail, and without knowing a great deal about the personal background of the prophet so that the prophets answers can be filtered through knowledge of the days of the prophecy itself, prophecy is difficult to interpret. Maybe that is as it should be. Maybe prophecy is meant to reveal itself only to those who gain a very high level of understanding at a particular time in history. The prophecy itself might just be clues that can only be understood by individuals who have attained certain knowledge already.

It would be ludicrous to assume that being created in the image of code means that God has a physical body and that we are cloned from the physical body of God. Different human races do exist, so that in itself eliminates the cloning interpretation. God is a spiritual being responsible for creating and maintaining the space of this entire universe. God created spiritual beings. The physical body is very much secondary. The dead rising to life could be nothing more than reincarnation, or it could be rejuvenation, it is not about exhuming corpses or trying to fit together the ashes of people who decided to be cremated rather than buried.

13

Extraterrestrials and UFOs

The previous chapters addressed the realities of large groups and individuals on planet earth, as well as the nature of future environmental change on the planet. Either this planet will be influenced by extraterrestrials in the future, or not. If not, then there is nothing to consider. If so, then all the challenges I have described could pale in comparison to the challenges presented by an alien race with advanced technology coveting this planet after it is "cleansed" so to speak from environmental change. Not matter how unlikely or wild that thought may be to most people, extreme situations need to be considered, especially when the odds cannot be honestly calculated. The true odds for winning the lottery for example, are very long odds that can be precisely calculated. Without even knowing the rules or playing field involved, it is not possible to calculate odds regarding extraterrestrial intervention or action against mankind. The reason extraterrestrials needs to be considered, and the reason some kind of public contingency plan needs to exist, is that consequences of doing nothing are so extreme. For example, what if one lone UFO with a gravity drive and a fusion weapon systems started firing on a U.S. city? Chances are the whole nation would panic, including the military, and who knows what that would lead to with thousands on nukes in silos across the nation. If extraterrestrials civilizations do exist, and a galactic terrorist high on some drug piloted that one ship, would it make sense for everyone to panic and launch dozens of nuclear weapons? That example is extremely unlikely, even if extraterrestrials do exist with the capability and desire to take some overt action towards earth. The focus of this chapter is an honest look at possibilities and available courses of action.

Not having the technology to observe this galaxy in detail, the odds of earth facing an extraterrestrial threat sometime in the near future could be anywhere from zero to a 100 percent certainty. The U.S. government has no public position regarding extraterrestrial threats or visitations, almost as if no one on the entire planet has ever considered it from a logical point of view much less presented contingencies.

Whatever the case may be, this chapter will present one scenario that is close to the truth about UFO's and extraterrestrials. This will be accomplished by eliminating any scenarios that can be disproved with confidence, and then focusing on a list of whatever possible scenarios remain. One scenario on that remaining list will be closest to the truth. That is one way to answer a question in the absence of direct observation data. Having done that, a contingency plan will presented to cover possible scenarios.

Considerable debate surrounds the subject of aliens and UFO's. There is a public project called the Search for Extraterrestrial Intelligence (SETI). SETI uses big dishes that listen for radio waves. Radio waves travel slower than light, and would not be a workable solution for communication between star systems. On earth, radio wave communications are generally used to communicate across distances in the range of a radio station broadcast. If a SETI receiver dish were setup 100 light years from earth, the chances of that same dish detecting anything besides noise coming from earth would be very close to zero. Add to that the fact that any radio wave communication is likely to use a form of encryption or compression, and it becomes clear that the technology used in the SETI project has almost no chance of finding a radio wave signal above the noise over interstellar distances. That does not mean the program should be cancelled, it just mean people should have realistic expectations regarding SETI project technology. Most likely, mankind will first have to learn how to travel to the stars before being able to proactively determine what civilizations if any exist in this galaxy.

There is no hard evidence that I am aware of on the subject of aliens and UFO's. If one is going to speculate it is a good idea to set down ground rules from a logical point of view. The question can be narrowed down to, "Assuming other civilizations exist with the means to get here, does it really matter?" That is the practical way of looking at the issue. If ET does not exist, or is not capable of getting here, or is to shy to do anything but hide upon arrival, then who really cares? Sure, it could become a good philosophical debate, but those scenarios have no effect on day-to-day life, and no contingency plan could be made for those cases.

Even looking at all recorded history, there is nothing obvious going on along the lines of the movie Independence Day. Independence Day was about a sudden alien invasion of earth. That narrows the entire field of extraterrestrials to either some kind of ongoing covert operation, or a surprise in the future. Alien conspiracy theories abound. If this is indeed a universe of life as I have argued, it is not unreasonable to assume that sentient life exists besides mankind, and that mankind could even share a spiritual link with other civilizations all being part of God's creation. Even if the physical form of sentient life is different on different planets, those alien individuals would still be a part of this universe created by the same God. The wide array of human emotions, and methods of reasoning, found in different people on earth should all apply to another civilization in this universe. That is a fair assumption, and one that needs to be made to go forward with this analysis of possible scenarios regarding extraterrestrials.

A wide range of motivations exists on this planet from torture and tyranny to kindness and caring. The desire to create and expand, freedom and slavery, all the elements of life, as we know it, could also exist looking at an alien civilization. Keeping that in mind, and knowing that at least we are not seeing any form of overt alien influence on this planet now, let us consider covert motivations.

On the covert operations list I doubt there is anything going on along the lines of "The Men In Black". There are so many people on this planet with cameras and video equipment it doesn't seem possible something like that could have escaped public detection. The idea that every single individual that saw something had his or her memory erased is just not possible. Granted, some people claim to have experienced UFO encounters, but no hard evidence of such encounters exists. Certainly such encounters would fall under the list of covert scenarios. The fact that so few people report such things out of billions of people on the planet, and the fact that no hard evidence exists for these encounters, leaves that scenario out of the picture in terms of contingency planning. Even if a few of these encounters are real, they have no effect on the general public, and there isn't anything constructive that could be done.

One popular book about alien conspiracy is called "The Gods Of Eden". This is an interesting book. "The Gods Of Eden" proposes that aliens start global conflicts covertly. There are a couple of problems with the logic behind this conspiracy proposal.

The U.S. would not have fought World War I, and World War II, if forces issuing forth from Germany did not seriously threaten the U.S. It is clear that the Nazi's were bent on global conquest, and that genocide was allowed to happen under the rule of Hitler. Hitler and the Nazis would either have had to be aliens, or hypnotized by aliens, for this to have been an alien scheme. Not only that, the people of Germany allowed these crimes to exist, some fearing to do anything granted. Dictators do not gain power without a large number of people giving up the right to self-government and supporting the dictator.

Looking back in history there is no shortage of examples documenting groups who have considered themselves superior to everyone else, then trying to prove it through military conquest. The fact is there is a mean streak in some people, and saying the aliens caused these individual actions is not a good defense in court even if that defense does manage to get some people out of jail and placed in an institutions instead.

Why would an alien race with destructive intent cause wars that only serve to increase military technology on the planet? If an alien race were bent on conquest, would it be wise to put greater weapons in the hands of the people on the planet one wanted to conquer? World War II did result in casualties, but the total number of casualties did not materially affect population growth. The U.S. lost ten times as many people in the civil war back when the total U.S. population was ten times lower.

"The Gods Of Eden" is a popular book on alien conspiracy theory so I present some conclusions drawn by that book. If covert aliens were involved in starting wars, then the motivations would have to be something other than those presented in "The Gods Of Eden." One result of all those wars was the fact that technology progressed rapidly. It does look like this planet is now in urgent need of technology. All of mankind would be in an even worse position today if the horse and buggy was still the pinnacle of technological advance.

I still do not buy the idea that ET descended and taught mankind how to build machine guns along with encouraging war. Although the rate of technological advance was rapid, mankind did go through a step-by-step development process. If ET wanted to introduce some technology, I would expect to have seen ray guns suddenly appear on the world stage. It could be that some bad people were just allowed to gain some power, or that people were just ignorant or misguided. Looking at it from a religious point of view, people could have misinterpreted the Bible or other religious scriptures resulting in war.

There is a Biblical verse that talks about slaying people that do not want to be part of the kingdom of God. As I mentioned earlier, I believe that the Bible clearly states that the individual faithful are not the wrath of God, and that the wrath talked about in Revelation is a natural force that needs to be prepared for. The creator of this universe could speak to a planet, sun, or galaxy just as easily, probably more easily, than speaking to an individual. It would be difficult for the prophets and saints who wrote the Bible to understand such things. There is the potential of great good in the Bible, warning all of mankind of events that were once beyond people's understanding. Unfortunately, that same lack of understanding, combined with negative intentions, could twist many things written in scripture towards destructive courses of action. Some Germans could have considered that the end of days was upon them, and that they were building a new civilization that the Jewish people did not want to be part of, justifying genocide. That reasoning would require more than one incorrect interpretation of scripture. This explanation makes more sense than the idea that aliens covertly started World War II.

Here is another alien covert ops scenario. Consider a Star Trek like cruise ship assigned to investigate earth with some kind of no interference prime directive. How would the local population be described? "Peace loving civilization, playing together nicely in the sand box and improving the environment." I don't think so. More likely the crew of the cruise ship would be glad they obeyed that prime directive on their approach, realizing that the combined military forces of the entire planet might do their best to shoot down that one ship. Even if the ship were to evade capture, there is a chance someone would panic and unleash nuclear weapons. Nuclear weapons are capable of turning this entire planet into a smoldering radioactive ball in space many times over. Certainly there would be no benefit in that for any civilization. Even an alien force bent on conquest would not want to win a planet turned into a radioactive cinder. So, the Star Trek prime directive idea is certainly a possible covert scenario. Once again, there is nothing practical that could be done, nor would there be any need for any kind of public information campaign, is this scenario is true.

Let's assume that habitable planets are rare in this galaxy. If that is true, Earth has value kept in good condition. A benevolent alien race might try to covertly introduce beneficial technology until mankind was open minded enough and reasonable enough to engage in diplomatic relations. The only was this could be done covertly would be using some form of telepathy, or spiritual inspiration. That is certainly a possible scenario. If this is true then people should be open minded and willing to learn new things. That is a good thing to be no matter what, so we have one possible scenario resulting in a positive contingency plan.

An alien force with hostile intent might also remain covert, but with different intentions. What would be the objectives of a hostile alien race? The way I see it, the only objective that would make sense for a violent race would be to take the planet while minimizing military loses and preserving the planet for future operations or colonization.

Put yourself in the shoes of an alien commander with a military space fleet given orders to secure earth. Let's say that this race has a Reptoid body form. Most reptiles are carnivorous and feed on mammals, so we will assume that the commander of this fleet has a low opinion of mammal life forms and is not likely to suddenly become benevolent towards mankind and question the orders he has been given. Pretending that you are this commander, what would you do? Your fleet and weapon systems are far superior to anything on the planet. You have an intelligence brief in front of you that says mankind possess thousands of nuclear weapons in the multi mega ton range capable of irradiating the entire surface of the planet for thousands of years. That would be bad, since your orders are to take the planet intact for colonization. The intelligence report also says that things are very unstable politically on the planet, and that some nations are ignorant enough to be blasting their own national territory with nuclear weapons for some reason or another. You wonder why a full-scale nuclear exchange has not begun already. Your geologist on board tells you that these nuclear blasts have accelerated the planets transition towards a magnetic pole shift, and that this transition is expected to rise to catastrophic levels for existing life on the planet in roughly ten years.

Being the commander of that alien force, and having a mission with an open-ended timetable, would you sit around and do nothing waiting for a reversal of the earth's magnetosphere? A pre-emptive strike on your part could easily result in nuclear war. On the other hand, your geologist predicts that all the earthquakes and volcanic events that the planet is rapidly approaching could bury some of those nuclear missiles in their silos. Your intelligence report also states that mankind is completely ignorant regarding the nature of planetary change, and is doing nothing to prepare for such an event, therefore the analysis indicates that 98 percent of mankind should be destroyed by natural causes in the not to distant future. The confusion created by such large-scale catastrophic events could paralyze governments, and provide a window of opportunity to neutralize ground based nuclear weapons. Even if some chemical rockets are launched, they can be shot out of the sky easily. As the commander of operation, you decide to wait and deal with whatever is left of mankind latter. Once natural events on the planet have taken their course, and the planet becomes hospitable once again, the time will be right for colonization. Maybe some of the humans will even survive as a primitive fire culture providing a local source of cattle.

That is a scary picture to paint, but not an unrealistic picture of how a hostile alien race would view mankind. Benevolent covert alien influences, and hostile covert alien influences, represent two ends of the spectrum of possibility. A really clever hostile alien force would try and influence the development of mankind by preventing new information from becoming public, especially information involving really advanced technology or geology.

This scenario is the second possible scenario that results in a practical contingency plan. That contingency plan would involve education combined with strengthening national defenses. Sometimes Democrats argue for defense cuts on the basis that the U.S. is already strong enough to handle any terrestrial threat. That is a valid argument. That argument assumes that the U.S. will never face an extraterrestrial threat. I have not provided any justification to prepare for an extraterrestrial threat. It all comes down to votes, and whether the majority of people want to play it on the safe side and defend the nation as best we can, or not.

There is a third possibility considering covert extraterrestrial operations. That possibility is an alien war in space, with one group wanting to help mankind and the other group wanting to destroy mankind. This is the most interesting scenario.

Stories about Atlantis exist. Maybe there is some truth in the idea that an advance civilization existed on this planet a long time ago, and that a natural catastrophe sunk that civilization under the ocean. The argument against that story is a question about the remains of that civilization. If poles shift do expand the diameter of earth, the remains of that civilization could be under hundreds of feet of rock that was once magma. A very advanced civilization with space travel technology could have abandoned this planet due to its tendency towards catastrophic change, in favor of a more stable planet elsewhere in the galaxy.

During the past 100 years technology has progressed at a torrid pace. Many great men have claimed that inspiration played a role in their research and discovery. More than one U.S. president has commented about "angels in the whirlwind". Have you ever had the feeling someone you cared for needed your help, even if they were not right in front of you telling you so? As strange as it may sound, there is no other explanation for that than telepathy, when you are right.

I believe that many technological advances we have experienced so far had an element of either divine intervention or telepathy involved. This divine intervention or telepathy from a benevolent race would be a principle method of helping guide mankind forward. The stories about Atlantis are that the people living on Atlantis were very much human, therefore the human race could have already expanded out into the stars.

The opposite of inspiration would be mind control. If mind control technology existed, that technology could be used by a hostile race. Mind control would be used to prevent people from seeing the truth, while putting plausible false information in place of the truth. False information would include ideas like an iron core model causing the magnetosphere, in the case of a respected researcher that the public looks up to. False information doesn't even have to be very plausible to work, some people just respond automatically to voices in their head. As bizarre as it may sound, technology may actually exist to inject voices and personality responses directly into the mind of an individual resulting in automated responses that are not really coming from the true heart and mind of the individual. There are only three possible reasons why people do the things they do. Those reasons include inspiration, an original idea, or an automatic response.

I am not going to try and convince people that mind control is possible. That would require building, operating, and demonstrating dangerous technology. People will just have to decide for themselves if this is something that should be a personal concern or not. I will just say that mind control technology is possible, and that most people would be very surprised to discover the nature of technology already deployed by certain elite groups to influence attitudes and behavior. It is not that difficult for individuals to rise above such manipulations. Rising above the problem is the correct solution. This entire universe is full of mind games, even in the scope of one on one relationships, so it is best to learn how to deal with it since it going to be an ongoing issue in life anyway. How does one rise above petty manipulations?

First and foremost, an individual must understand why they are doing what they are doing when it comes to making decisions. Most automated responses should be avoided. That does not include physical responses driven by the bodies built in defense mechanisms. For example, accidentally touching a very hot surface the body should react and move the hand away. The automated responses that should be avoided involve communications that are delivered, and decisions that are made, which do not take into account the present situation and future consequences. Every communication, and every decision, especially deciding not to decide or not to look at someone or something honestly, does have some future consequence. Sometimes that consequence is not important, and sometimes it is. Evaluating that future consequence is the way to avoid being manipulated and trapped by automated responses.

The difference between inspiration and mind control is that one produces a positive result for the individual and the other produces a negative result.

Living life, it is not practical to consider all the communications and decisions ones makes in a day. I doubt it is even practical to consider even 5 percent of those communications and decisions. The solution is to do an after action review at the end of the day or at the end of the week. If a pattern is noticed linking certain actions to a negative result, or to results that do not make towards progress achieving personal dreams, try doing something else in the future when that kind of situation arises that involves a logical plan or a well-considered response. Doing this, one can gradually improve and know that one is living life rather than just being manipulated. In the game of mind control the word maybe is like rolling out the heavy artillery. There is nothing in this universe more paralyzing than a maybe. Maybe is like locking away part of an individual's mind in prison then throwing away the key. Given enough maybes in life an individual becomes completely incapable of original thought or future planning. An individual with some maybes in their life would feel better changing all those maybes to a yes or a no. If a situation is to complex to resolve immediately, then list the possible scenarios including contingency plans. Getting rid of those maybes, and changing them to a yes or a no, it is still possible to change ones mind latter. At least the maybe is not sitting in ones mind the whole time drawing a continuous flow of personal attention.

That covers all the scenarios that I consider plausible regarding covert influences, including the possibility of mind control. I have discounted overt alien influences. Could it be that overt operations exist that are being cover up? There is enough information regarding area 51 to suspect a government cover up in that case. The argument is that an alien craft crashed, and the electronics technology including the transistor arrived as a result of investigating that crash.

Whether that is true or not has very little bearing on the future. Maybe transistor technology came out of a crashed alien craft, maybe not. At this point people understand computer technology. Maybe gravitational drives have been reverse engineered by the department of defense, maybe not. I would prefer working from scratch on something like that, using the models I have presented.

The Art Bell show hosted an individual who talked about trying to reverse engineer the propulsion system of an alien craft without success. The radio talk was very detailed, and it did all fit together. A smaller craft would contain a core energy field that would simply float down into the planet's core once the craft's containment field was compromised. I would speculate that the equipment required to generate a primary core field would not be present onboard a small craft. That kind of energy field would not influence the magnetosphere. It takes something like a large hydrogen bomb to have a significant impact on something as large as the planet. I doubt that technology exists anywhere that is more destructive than hydrogen bombs. So, I would not be surprised if some alien artifact or damaged craft was stashed away some place.

Mankind will not know the truth of the extraterrestrial situation until one of two things happen. Either aliens become overt, or technology is developed to take a very close look at this galaxy. One day, probably sooner than most people expect, mankind will reach for the stars. The nation needs something to strive for. A bright future that people can understand and make progress towards achieving.

So far I have presented some very general scenarios regarding aliens and UFO's. Surfing around on the net, one can find a great deal of entertaining material on the subject of extraterrestrials. Since technology does not exist to see, visit, or communicate outside this solar system, stories told by spiritualists represent the only specific information available be it true or false. I searched many different sites and found similar stories from unrelated individuals. So, a common theme does exist. That story is worth telling.

The best evidence would be a big ship landing for inspection, or a ship sitting in close orbit that the shuttle could go visit and inspect once the space program is resumed. Having narrowed things down to covert ops that cannot be expected. If two alien groups engaged in conflict in this solar system, one would expect some kind of military stealth technology to be used. If a large ship was destroyed within this solar system the explosion might be big enough to see from earth. Most likely that would look like a shooting star. Meteor showers are not uncommon these days. In any case, I would not bet on earth technology being capable of seeing anything in detail that a civilization with space travel technology did not want to be seen.

The comet Shoemaker Levy 9 was discovered in March of 1993. Shoemaker Levy 9 must have been a sizable comet. No one knows what it looked like originally because it was suddenly discovered broken up into 21 pieces all being pulled into the gravitational well of Jupiter. The pieces themselves were not that difficult to spot. Why it was not seen as a bigger single piece is still a mystery. Scientists have determined that it was all one object originally because the pieces were strung out like pearls close together.

Comets orbit the sun in this solar system. Shoemaker Levy 9 was found to be in orbit around Jupiter. The chances of Jupiter capturing a comet into its orbit have been estimated a 1 in a billion. Archived images of Jupiter exist prior to the initial discovery in 1993 and none of these images show the comet. All that is known for certain is that it was once a larger object that broke into pieces. Solid asteroids do not suddenly break into pieces, nor have comets previously exhibited this behavior.

Prior to the collision with Jupiter NASA got a very good look at the comet pieces through the Hubble space telescope. Mass spectrum data was gathered by NASA's Faint Object Spectrograph (FOS). A comet is supposed to be mostly ice. One would expect hydroxyl ions (OH) in the comet tail. The tail should also replenish itself evenly as gases escape. OH ions were not observed in the tail. The comet tail did not replenish itself consistently. A spike in magnesium was observed at one point studying the spectrograph data.

During July of 1994 the 21 pieces of Shoemaker Levy 9 collided with Jupiter. NASA's FOS detected silicon, magnesium, and iron at the impact area. Some pieces exploded while entering Jupiter's atmosphere. Silicon, magnesium, and iron are not part of Jupiter's atmosphere so these elements must have come from the comet. The explosions in the atmosphere of Jupiter were much larger than expected. Shoemaker Levy 9 was something other than a big ball of ice.

Now lets begin the extraterrestrial soap opera story. Some spiritualists say that Shoemaker Levy 9 was actually a large Reptoid colonization ship carrying hundreds of thousands of Reptoids in a suspended state ready to colonize earth. The ship orbited Jupiter, assuming that this solar system was faithfully guarded by another alien race called the Grays, a race created by the Reptoids using genetic engineering. The Grays have the big eyes and small mouth commonly depicted on T-Shirts, in movies, etc.

There are a variety of races that exist in this galaxy. One race is very similar to the human race, fair haired with a sort of Nordic look. In the distant past various human races from different parts of the galaxy visited this planet and created human societies. Out in the galaxy, the most advanced race of human beings are believed to be the Pleiadians. The Pleiadians are named by the star system they are thought to have originated from. The Pleiadians have a very long history in this galaxy, dating back before the arrival of the Reptoid race in this galaxy. The Reptoids are considered an invading force within this galaxy.

The Pleiadians are considered to exist as much in a spiritual sense as in a physical sense. The Pleiadian philosophy is one of love, selfless service, and dedication to stewardship in the name of God, the creator of this universe. Some Pleiadians simply exist in a spiritual form, helping guide the human races. The Pleiadians have lived on their own planets at peace among themselves for hundreds of thousands of years. At the same time, the Pleiadians have fought against other races, including the Reptoids, which would disturb that peace.

The Reptoid race looks similar to the creatures depicted in the television series "V". "V" was a series about earth being taken over by this Reptoid race. The Reptoids are thought to be very advanced, and capable of inter-galactic space travel. The Reptoid race recognizes the spiritual nature of existence, understanding exactly how the spirit transcends the physical body. At the same time, the Reptoid race is self-serving. Reptoids could be considered spiritual atheists recognizing the soul but denying the existence of God. Reptoids are carnivorous, and would consider backwards humans little more than cattle.

The goal of the Reptoids is conquest and colonization in this galaxy. The Grays, being not very threatening in person, are used for reconnaissance, security, and infiltration of societies prior to large-scale military action.

Presently there is a full-scale galactic war in progress that has been evenly balanced for some time. This war would involve various human races, led by the Pleiadians, versus Reptoids and Grays. Some attention has been focused on earth.

If the war were to start going against the Reptoids, the Grays would have to make some contingency plans. Some Grays betrayed their Reptoid masters in this system, opening a security breach and allowing the destruction of a large Reptoid colonization ship.

That concludes one story that emerges visiting various sites on the Internet dedicated to alien races, UFO's, etc. To be on the safe side, I would support building advanced military defenses unmatched by anything else in the galaxy. Sometimes freedom has to be protected. There are two many unknowns surrounding the subject of extraterrestrials to dismiss the subject out of hand. If there is something going on the public should know what the situation is.

14

Conclusion

This book represents a leap forward for mankind in the fields of health, energy, physics, chemistry, and spirituality. An advance in any one of these fields is selected each year by the Nobel foundation based on which advance delivers the greatest real benefit to mankind. Previous winners of the Nobel Prize, a few university department heads around the world, and the Nobel foundation itself are the only individuals who may sponsor someone to compete for the Nobel Prize. A friend and business partner of mine who contracted two previous winners of the Nobel Prize in physics asked me when I would be receiving the award. A particular technology project my partner was involved in could not proceed constructively without a greater understanding of the energy involved. That understanding did not exist prior to this publication. The support and compliments of partners and friends is more important to me personally than a formal prize. Public recognition is more important to the public than an individual. Personally, I do not need formal recognition, nor do my close friends or partners. However, the average public individual who is not capable of making their own decisions or evaluations regarding technology relies on the media and formal 3rd party sources to make decisions for them. This exact tendency is one reason why the general public is so very easy to manipulate and control. That is just the way things are. That statement is not intended to blame anyone for being ignorant or for exercising control. Public recognition does validate new ideas and makes those ideas part of public awareness. From a political point of view, that is important.

Out of the hundred or so people who have won the Nobel Prize in physics, Einstein stands head and shoulders above the rest. Einstein was still criticized when his work was originally released, since he had no way of proving it to people. In time, after nuclear bombs were successfully detonated and people tested Einstein's equations regarding time dilation, Einstein's work became a part of public awareness. This book does present more than ideas. A variety of lower energy products are available for testing. These products do have a measurable, predictable effect on water that can only be explained by stable spheres of energy acting on the water in the absence of a chemical reaction. The health benefits of these products also support the models presented in this book.

It is tempting to build a fusion blaster and ask anyone who doubts the technology to stand down range for a test. That would be quite extreme though, I am sure a video would be just as convincing. In time, I am sure such a test will be done for the U.S. military, with generals standing up range of course behind protective shielding. For now, the products that are currently available will have to serve as a proof of concept. I look forward to meeting some people who have rejuvenated to the degree that I have.

Anyone with a pH meter, or finger for that matter, can test a product like the Crystal Beads. The Crystal Beads will soften water and raise the pH of water without any lose of mass. To do this test, a professional pH meter is required with a range going from at least neutral up to 10.5. Cheap pH strips do not have a great enough range or resolution. A pH meter costs about $100. A meter for water hardness costs more, but most people will be able to feel the change when starting with typical hard tap water. Use the Crystal Beads on a few gallons of water first to make sure they are completely rinsed off. Some Crystal Dust may be on the beads small enough to fall through the mesh bag. Weigh the Crystal Beads dry before converting a gallon of water, dry them, and then weigh them again. An accurate scale would also be needed to do this test. If you are not a chemist, write down the starting pH of the water and the ending pH then ask a chemistry major to calculate the weight of the chemical reactant that would be needed to create that change in pH.

Hear is a more subjective and less expensive test. Convert a gallon of distilled water with the Crystal Beads and put it in the refrigerator. See if your body doesn't go running to that refrigerator for water more so than usual.

Most of the new things I have presented are verifiable either using existing technology or by personal experience. Here is a summary of the major topics that have been presented.

The nature of creation, and the development of planets, is now understood. Earth is approaching a major electromagnetic reversal cycle that will bring with it catastrophic change. Once this cycle is complete, earth will return to a stable electromagnetic dipole mode conducive to the evolution of life as the earth is today. As it stands in 2003 virtually none of mankind is prepared to endure the extreme environment that will accompany change. Survival system technology exists that could be used to provide for people in such an extreme environment, however that technology is sensitive and cannot be released to the public without major political change and the approval of the executive branch. Therefore, a grass roots political movement is needed to encourage positive change.

The economy has been clearly defined as the efforts of individuals contributing to Gross National Product. An economy that lacks vision and leadership has a tendency to stagnate. Private enterprises is still the best way to go, but the expansion of enterprise does need to be the focus of team work between the public and private groups. Lacking a vision of the future, the U.S. has painted itself into a corner building the same old houses, cars, and appliances year after year to the point of market saturation all in the context of centralized energy and food distribution system. Continuing down that path much longer will result in economic collapse. Economic change is needed, and the goals of positive change need to be clearly understood and supported by the majority of the public. Change will not be easy to achieve, considering that a large number of powerful special interest groups exist that would tend towards keeping things the way they are today.

Looking at the changes in the weather lately, most people realize that some kind of change is happening on a global scale. The greatest challenge that people will face talking about the ideas in this book is a denial of understanding. This book is true, and people do have an innate ability to recognize truth. Even so, many people second guess their own intuition and cling to ideas of the past. This tendency results in individuals failing to take a definitive stance. A maybe is much worse than a contrary argument. It is not possible to work with someone who will not even argue with you. A person stuck in a maybe on a particular issue is like a mind frozen in time and perched on the edge of complete nothingness and chaos.

The possibility of the existence of extraterrestrials interested in earth has been considered. Not knowing what the actual situation is, a couple scenarios have been identified that imply contingency plans that can be acted on today. The first contingency plan is an individual response, being open minded and willing to experience positive spiritual inspiration. The second contingency plan is to continue the expansion of the U.S. department of defense just in case the department of defense is needed to counter an extraterrestrial threat. The decision to pursue or not pursue this second contingency plan will be up to congress and voters. The U.S. has agreed to certain international treaties like not bringing weapons into space. If a decision were made to expand the role of the U.S. department of defense some treaties would have to be either reconsidered or abandoned.

The future has been defined as the actions and intentions of billions of people with free will combined with a certain amount of divine destiny. In the past, Biblical prophecy has proven true on many occasions. Destiny has been defined not as something that ones goes searching for, but rather that which will come to be no matter what. With this in mind individuals should not worry about destiny. Individuals should simply live their lives as best they can. It may be that the end goal of destiny for mankind is that mankind gains a high enough level of understanding to shape destiny and bring about an ideal civilization.

Eternal youth and rejuvenation are now possible. Complete rejuvenation takes many months if not years, depending on how much of the body needs to be rebuilt, it also requires dedication, a quality diet, time for rest and energetic healing, and a significant quantity of crystal products. It is much easier to retain youth than it is to rejuvenate the effects of aging; that rejuvenation is even possible is as close to miraculous as things get. Even with aging and death planetary population growth has been exponential over the past couple hundred years. In many parts of the world populations are pushing the limits that can be supported by the land. Living such short life spans, most people have not been particularly concerned about things like that. Even so, population expansion is something that will need to be addressed especially if everyone wants to enjoy vibrant health for hundreds of years combined with the joy of raising new families. In the grand scheme of things, it really doesn't take that long for an expanding civilization to fill up planets. Moving billions of people to a newly discovered planet would be a quite a challenge even for the most advanced civilization, and even that solution would only be temporary. I am quite sure that a great civilization existed in this galaxy a very long time ago spanning many planets with rejuvenation technology and faster than light travel. This same expansion issue resulted in the downfall of that civilization. Failing to recognize the scope of that expansion challenge, and failing to implement a democratic policy to manage expansion, were the main factors contributing to the destruction of that civilization.

Detailed answers to every question and issue involving life and this universe would fill millions of sets of encyclopedias. So, why is this book called "Answers For Everything"? Is it false advertising? Not exactly.

The text on the back cover of the book says that knowledge is a pyramid. The pinnacle of that pyramid of knowledge represents ultimate truth and simplicity from which all other knowledge can be derived. The knowledge that would fill those million sets of encyclopedias are essentially derivations. Here is an example. My friend Andy read this book and asked me why the book was called "Answers For Everything". Andy was hoping to learn how to fix his alternator, and as it turns out his alternator still wasn't working.

The technology of building alternators is based on Maxwell's laws. The models that I have presented are a generalization of Maxwell's laws, and really are not needed to build alternators. If anyone wants to replace that alternator with fusion generator, then they would have a tough time doing without making use of knowledge at the top of the pyramid related to this universe. So, how is Andy going to learn how to fix his alternator? If Andy did very well in his math classes in high school, he would have the prerequisites to begin his first year of college studying electrical engineering. During that first year, Andy would learn about Maxwell's laws in detail and learn how to derive those laws from observations people have made of the universe. During his next three years of college, Andy would learn how design simple electronic circuits. After that, Andy would have the knowledge necessary to understand and trouble shoot his alternator. Not wanting to do that, most people just go out and buy a new alternator. The study of electrical engineering represents moving down the pyramid of knowledge. As one moves down that pyramid in any field of endeavor, things become more and more complex.

Alternator technology is extremely simple for an electrical engineer. Most electrical engineers go to work for a company after graduation and are then assigned real projects. For example, and an electrical engineer might be asked to design a new hand held receiver for a Global Positioning System (GPS). The engineer's boss would add certain constraints to the project after describing what the receiver needed to do. Those constraints would include things like size constraints for the receiver, cost constraints for the product, certain aesthetic requirements to make the product marketable, and a project dead line. The electrical engineer right out of college who has only learned how to design simple circuits would then be over whelmed. The new engineer would seek the help of a real engineer with 10 or 15 years of experiences designing GPS receivers. Those 10 or 15 years of experience in a specialized field represent an enormous amount of detail knowledge farther down on the pyramid of knowledge related to electrical engineering. Even so, it is all related to spheres of energy called electrons moving about and interacting. This book is called answers for everything because it will be found that all fields of knowledge related to life and creation in this universe involves fields of energy, or the interactions between fields of energy.

One might argue that there is more to life than energy. Fields of energy produce a core singularity inside which the laws of this universe no longer exist. Outside that singularity, the laws of this universe do exist, and those laws influence the singularity. There is more to a singularity than the classic galactic black hole. Derivations of the models I have presented will show that a singularity can be of any size and of any strength, even existing in the same space as anything else in this universe. The quality of imagination, awareness, and memory are all associated with a singularity. So, the models I have presented are indeed a big umbrella.

The pinnacle of knowledge represents both the end of confusion, and the beginning of an unlimited era of new discovery. Looking out from the pinnacle of knowledge everything is possible. The back of the one dollar bill has a good picture of the pinnacle of knowledge in the great seal of the United States. I look forward to reading some of those new books as people start filling up those million sets of encyclopedias.

To find the answers to your own questions look as a child looks, with trust and faith, and without fear or preconceived notions. The opposite of living a life of joy and seeing through the eyes of a child is living a life of fear. Fear takes on many faces from a fear of financial loss, to a fear of physical injury or death. Why anyone would fear embarrassment is beyond me, but apparently some people do. The problem with fearing loss of any kind is that the fear itself is directly related to a decision that one cannot simply create again that which was let go of. That decision, deciding one cannot create again, is a decision for the void and chaos against life itself. There is an old saying that an entrepreneur has to go through 3 bankruptcies before succeeding in business. Piling up a bunch of nonessential stuff in life is like an athlete picking up bricks every 100 yards in a five-mile race. That athlete certainly will not win, and will probably not be having much fun in the race after a few laps around the track carrying all those bricks. The challenges of life teach the value of freedom. Challenge is a learning experience that requires more than one try until something finally works. There is no such that as failure in anything as long as one is still willing to try.

Finding your own answers may take a lot of looking. The search may lead down a dozen paths, many of which may not even appear related to the original question at first. Some paths may be hard to travel. Be still for a moment then seek and you shall find, knock and the door shall be opened to you.

Thank you for reading.

Index

0-595-28071-4